COHOMOLOGY OPERATIONS and APPLICATIONS in HOMOTOPY THEORY

Robert E. Mosher
Professor Emeritus
California State College at Long Beach

Martin C. Tangora
Professor Emeritus
University of Illinois at Chicago

Dover Publications, Inc., Mineola, New York

Bibliographical Note

This Dover edition, first published in 2008, is an unabridged republication of the work originally published by Harper & Row, Publishers, Inc., New York, in 1968.

Library of Congress Cataloging-in-Publication Data

Mosher, Robert E.
Cohomology operations and applications in homotopy theory / Robert E. Mosher, Martin C. Tangora.—Dover ed.
p. cm.
Originally published: New York : Harper & Row Publishers, 1968.
Includes bibliographical references and index.
ISBN-13: 978-0-486-46664-4
ISBN-10: 0-486-46664-7
1. Homology theory. 2. Homotopy theory. I. Tangora, Martin C. II. Title.

QA612.M68 2008
514'.23—dc22

2008009891

Printed in Canada
46664704 2025
www.doverpublications.com

CONTENTS

PREFACE

In the past two decades, cohomology operations have been the center of a major area of activity in algebraic topology. This technique for supplementing and enriching the algebraic structure of the cohomology ring led to important progress, both in general homotopy theory and in specific geometric applications. For both theoretical and practical reasons, much analysis has been made of the formal properties of families of operations.

We focus attention on the single most important sort of operations, the Steenrod squares. We construct these operations, prove their major properties, and give numerous applications.

In the development of these applications, we treat various techniques of homotopy theory, notably those useful for computation. In the later chapters, special emphasis is placed on calculations in the stable range. Here we make detailed computation in the stable homotopy of spheres and, in Chapter 18, develop the tool currently found most effective for stable calculations.

This book is intended for advanced graduate students who are well versed in cohomology theory and have some acquaintance with homotopy groups. It is based on notes by the second-named author from lectures aimed at such students and given at Northwestern University by the first-named author. It attempts to give the student a thorough understanding of the cohomological methods and their history.

Our choice of treatments has not generally been based on elegance but rather, we feel, on ease of comprehension; this book is primarily for the student. One can find in the literature a smoother definition of the squaring operations, a slick exposition of spectral sequences, or an axiomatic approach to secondary operations. These methods, important as they are, do not seem entirely appropriate for the student's first view of the subject. They rarely were the discoverer's first view either.

The book makes no pretense to being self-contained. To fill in complete details at every stage would be to compromise the goals of the book, which aims to give an overview of the theory and practice of the cohomological method and to attract students into the field. We have collected many results which until now were scattered through the literature, and some have been proved in detail while others have only been stated (with reference to sources). We have used various criteria in deciding when or when not to give details, but of course the final decision ultimately rests on the personal taste of the authors.

We wish to thank Professors A. H. Copeland, Jr., Daniel S. Kahn, and R. C. O'Neill for their suggestions and the National Science Foundation for partial support. And in particular we thank Professor F. P. Peterson, not only for his suggestions but also for his lectures at the Massachusetts Institute of Technology, to which the authors acknowledge a substantial debt.

ROBERT E. MOSHER
MARTIN C. TANGORA

CHAPTER 1

INTRODUCTION TO COHOMOLOGY OPERATIONS

It has long been known that the sphere S^{n-1} is not a retract of the closed cell E^n of which it forms the boundary. Such a retraction r would, when composed with the inclusion map i, give the identity map of S^{n-1}. Passing to the reduced homology groups, we would have a factorization through $\tilde{H}_*(E^n)$ of the identity map of $\tilde{H}_*(S^{n-1})$: $\tilde{H}_*(S^{n-1}) \xrightarrow{i_*} \tilde{H}_*(E^n) \xrightarrow{r_*} \tilde{H}_*(S^{n-1})$. This is impossible because the reduced homology of the cell is zero whereas that of the sphere is not: $\tilde{H}_{n-1}(S^{n-1})$ is an infinite cyclic group.

This argument illustrates one of the fundamental methods of algebraic topology: the non-existence of a topological construct (the retracting map) is proved by showing that its existence would entail the existence of an algebraic construct (the induced map of the homology groups) which is more readily seen to be impossible.

We consider another, more delicate problem. Let $P(n)$ denote real projective n-space ($n > 1$). It is known that the fundamental group $\pi_1(P(n))$ is Z_2, that is, the cyclic group of order 2, independent of n. Question: For $n > m > 1$, does there exist a map $f: P(n) \to P(m)$ inducing an isomorphism of the fundamental group? Again the non-existence of the map can be demonstrated by considering an induced algebraic map. We consider the cohomology *ring* of $P(n)$ with Z_2 coefficients. This ring is known to be a truncated polynomial ring on a one-dimensional cohomology class a_n, truncated by the relation $(a_n)^{n+1} = 0$. If there existed a map f with the required property, then the induced map in cohomology would have to satisfy $f^*(a_m) = a_n$. (This follows from the Hurewicz theorem and the universal coefficient theorem for $H^1(P(n); Z_2)$.) Because f^* must be a *ring* homomorphism, this is impossible, for we would have $(a_n)^n \neq 0 = f^*(0) = f^*((a_m)^n)$ when $n > m$.

This example illustrates the advantage of cohomology over homology because of the additional algebraic structure given by the cup product. Our first objective in this book will be to develop a much more extensive algebraic structure in cohomology; the cup product will be supplemented, or overwhelmed, by an infinite family of operations. Every topological map respects not only the homology groups and the cohomology ring but all the structure afforded by these new operations.

We now make precise what we mean by a cohomology operation.

COHOMOLOGY OPERATIONS AND $K(\pi,n)$ SPACES

Definition

A *cohomology operation* of type $(\pi,n; G,m)$ is a family of functions $\theta_X: H^n(X;\pi) \to H^m(X;G)$, one for each space X, satisfying the naturality condition $f^*\theta_Y = \theta_X f^*$ for any map $f: X \to Y$.

EXAMPLE: The cup-product square, $u \to u^2$, gives for each n and each ring π an operation $H^n(X;\pi) \to H^{2n}(X;\pi)$. Note that this is generally not a homomorphism.

We will denote by $\mathcal{O}(\pi,n; G,m)$ the set of cohomology operations of type $(\pi,n; G,m)$. In Theorem 2 we will establish a fundamental result on the classification of such operations in terms of the cohomology of certain spaces, which we now introduce.

Definition

We denote by $K(\pi,n)$ any "nice" space which has only one non-trivial homotopy group, namely, $\pi_n(K(\pi,n)) = \pi$.

By a "nice" space we mean a space having the homotopy type of a CW-complex. We will usually assume that the spaces under consideration are of this type, without making explicit mention.

Recall that the *Hurewicz homomorphism* $h: \pi_i(X) \to H_i(X)$ is defined for any X and any i by choosing a generator u of $H_i(S^i)$ and putting $h: [f] \to f_*(u)$ where $f: S^i \to X$.

A space X is said to be *n-connected* if $\pi_i(X)$ is trivial for all $i \leq n$. The *Hurewicz theorem* states that, if X is $(n-1)$-connected, then the Hurewicz homomorphism is an isomorphism in dimensions $i \leq n$ and is still an epimorphism in dimension $n+1$. Thus the first non-trivial homotopy group of X and the first non-trivial homology group occur in the same dimension and are isomorphic under h. The statement of the theorem must be modified for $n = 1$; in this case the epimorphism $h: \pi_1(X) \to H_1(X)$ has as kernel the commutator subgroup of $\pi_1(X)$—and we do not claim that h maps $\pi_2(X)$ onto $H_2(X)$.

The universal coefficient theorem for cohomology gives an exact sequence

$$0 \to \text{Ext}\,(H_{n-1}(X),\pi) \to H^n(X;\pi) \to \text{Hom}\,(H_n(X),\pi) \to 0$$

for any space X. If X is $(n-1)$-connected, this becomes

$$H^n(X;\pi) \approx \text{Hom}\,(H_n(X),\pi)$$

since the Ext term becomes Ext $(0,\pi) = 0$. Now if $\pi = \pi_n(X)$, then the group Hom $(H_n(X),\pi_n(X))$ contains h^{-1}, the inverse of the Hurewicz homomorphism (which is an isomorphism here).

Definition

Let X be $(n-1)$-connected. The *fundamental class* of X is the cohomology class $\iota \in H^n(X;\pi_n(X))$ which corresponds to h^{-1} under the above isomorphism. This class may be denoted ι_X or ι_n, to stress its provenance or its dimension.

In particular, $K(\pi,n)$ has a fundamental class $\iota_n \in H^n(K(\pi,n);\pi)$.

Later in this chapter we will prove the following theorem.

Theorem 1

There is a one-to-one correspondence $[X, K(\pi,n)] \leftrightarrow H^n(X;\pi)$, given by $[f] \leftrightarrow f^*(\iota_n)$. (Here the square brackets denote homotopy classes, as usual.)

We draw two important corollaries.

Corollary 1

There is a one-to-one correspondence

$$[K(\pi,n), K(\pi',n)] \leftrightarrow \text{Hom}\,(\pi,\pi')$$

This follows from Theorem 1, the universal coefficient theorem, and the Hurewicz theorem:

$$\begin{aligned}[K(\pi,n), K(\pi',n)] &\leftrightarrow H^n(K(\pi,n);\pi') \\ &\approx \text{Hom}\,(H_n(K(\pi,n)),\pi') = \text{Hom}\,(\pi,\pi')\end{aligned}$$

Corollary 2

The homotopy type of $K(\pi,n)$ is determined by π and n. Moreover, the identity map of π determines (up to homotopy) a canonical homotopy equivalence between any two "copies" of $K(\pi,n)$.

Recall that $K(\pi,n)$ is assumed to have the homotopy type of a CW-complex. By Corollary 1, any isomorphism $\pi \approx \pi'$ can be realized by a map $K(\pi,n) \to K(\pi',n)$. Since all the other homotopy groups are trivial, it follows by J. H. C. Whitehead's theorem that this map is a homotopy equivalence. This proves Corollary 2.

Notation

We write $H^m(\pi,n; G)$ for $H^m(K(\pi,n); G)$. This is justified by Corollary 2.

We come to the classification theorem. Let θ be an operation of type $(\pi,n; G,m)$. Since $\iota_n \in H^n(\pi,n; \pi)$, we have $\theta(\iota_n) \in H^m(\pi,n; G)$.

Theorem 2

There is a one-to-one correspondence

$$\mathcal{O}(\pi,n; G,m) \leftrightarrow H^m(\pi,n; G)$$

given by $\theta \leftrightarrow \theta(\iota_n)$.

PROOF: We will show that the function $\theta \leftrightarrow \theta(\iota_n)$ has a two-sided inverse.

Let $\varphi \in H^m(\pi,n; G)$. We define an operation of type $(\pi,n; G,m)$, which we also denote by φ, as follows: for $u \in H^n(X; \pi)$, put $\varphi(u) = f^*(\varphi) \in H^m(X; G)$ where f is a map $X \to K(\pi,n)$ such that $[f]$ corresponds to u under Theorem 1.

We now have functions in both directions between $\mathcal{O}(\pi,n; G,m)$ and $H^m(\pi,n; G)$, and it is easy to see that the composition in either sense is the identity transformation. In fact, if $X = K(\pi,n)$ and $u = \iota_n$, then f must be homotopic to the identity, and so $\varphi(\iota_n) = f^*(\varphi) = \varphi$. On the other hand, if $\varphi = \theta(\iota_n)$, then the operation φ is equal to θ, since $\varphi(u) = f^*(\varphi) = f^*(\theta(\iota_n)) = \theta(f^*(\iota_n)) = \theta(u)$. This completes the proof.

Usually we will use the same symbol to denote both an operation and the corresponding cohomology class, as we have already done for φ in the proof above.

Corollary 3

There is a one-to-one correspondence

$$\mathcal{O}(\pi,n; G,m) \leftrightarrow [K(\pi,n), K(G,m)]$$

This follows immediately from Theorems 1 and 2.

Theorem 2 reduces the problem of finding all cohomology operations (of the type under consideration) to the computation of the cohomology of $K(\pi,n)$ spaces. Of course $K(\pi,n)$ spaces are not geometrically simple; they are generally infinite-dimensional, with the striking exception of the circle

$K(Z,1)$. Computation of their cohomology will be one of our major objectives. But first we show that $K(\pi,n)$ spaces exist.

Construction

Let π be an abelian group and n an integer, $n \geq 2$. We construct a CW-complex $K(\pi,n)$.

To begin, let $0 \to R \to F \to \pi \to 0$ be a free abelian resolution of π, and let $\{a_i\}$ and $\{b_j\}$ form bases from R and F, respectively, where i and j run through index sets I and J. Let K be the wedge (one-point union) of n-spheres S_j^n, $j \in J$. For each a_i, take an $(n+1)$-cell e_i and attach it to K by a map $f_i: \dot{e}_i \to K$, where $[f_i] = a_i \in \pi_n(K) = F$ and $\dot{e}_i$ denotes the boundary of e_i. Then the resulting space X, formed from the disjoint union of K and the e_i by identifying each point in $\dot{e}_i$ with its image under f_i, is of dimension at most $n+1$, has $\pi_n(X) = \pi$, and is $(n-1)$-connected (by the Hurewicz theorem).

It remains to kill the higher homotopy groups. This is done by transfinite induction, using the following lemma.

Lemma 1

Attach an $(m+1)$-cell to the CW-complex Y by a map $f: S^m \to Y$, and call the resulting complex Z. Then the injection $j: Y \to Z$ induces isomorphisms $j_\# : \pi_i(Y) \approx \pi_i(Z)$, $i < m$; in dimension m, $j_\#$ is an epimorphism, which takes the subgroup generated by $[f]$ to zero.

PROOF: The isomorphism for $i < m$ follows from the cellular approximation theorem, since Y and Z have the same m-skeleton. If i,j,k are the injections, the diagram below is commutative.

$$\begin{array}{ccc} S^m = \dot{e}^{m+1} & \xrightarrow{\ i\ } & e^{m+1} \\ \downarrow f & & \downarrow k \\ Y & \xrightarrow{\ j\ } & Z = Y \cup_f e^{m+1} \end{array}$$

Passing to the induced diagram in homotopy, $j_\# f_\# = k_\# i_\# = 0$ (since $i_\# = 0$). If $[1]$ denotes the generator of $\pi_m(S^m) = Z$, then $j_\#[f] = j_\# f_\#[1] = 0$. Thus $j_\#$, being a homomorphism, takes $[f]$ and the subgroup generated by it to zero. Finally, the assertion that $j_\#$ is onto $\pi_m(Z)$ follows from cellular approximation, since any map of S^m into Z can be deformed into the m-skeleton $Z^m = Y$.

This proves the lemma and completes the construction.

The balance of this chapter is devoted to a survey of obstruction theory and the proof of Theorem 1.

THE MACHINERY OF OBSTRUCTION THEORY

We now recall some machinery of obstruction theory which will be needed to prove Theorem 1.

Let Y be a space, and assume for convenience that Y is *n-simple* for every n, that is, the action of $\pi_1(Y)$ on $\pi_n(Y)$ is trivial for every n. Under this hypothesis we can forget about base points for homotopy groups, and any map $f: S^n \to Y$ determines an element of $\pi_n(Y)$.

Let B be a complex and A a subcomplex. Write X^n for $A \cup B^n$, where as usual B^n denotes the n-skeleton of B. Let σ be an $(n+1)$-cell of B which is not in A. Let $g = g_\sigma$ be the attaching map $\dot{\sigma} = S^n \to X^n \subset B$.

Given a map $f: X^n \to Y$, denote by $c(f)$ the cochain in $C^{n+1}(B,A;\pi_n(Y))$ given by $c(f): \sigma \to [f \circ g_\sigma]$. Then it is clear that f may be extended over $X^n \cup_{g_\sigma} \sigma$ if and only if $f \circ g_\sigma$ is null-homotopic, that is, if and only if $c(f)(\sigma) = 0$, and therefore that f can be extended over $X^{n+1} = A \cup B^{n+1}$ if and only if the cochain $c(f)$ is the zero cochain. It is a theorem of obstruction theory that $c(f)$ is a cocycle. It is called the *obstruction cocycle* or "the obstruction to extending f over B^{n+1}."

There are two immediate applications. First, any map of an n-dimensional complex K into an n-connected space X is null-homotopic.

PROOF: Take $(B,A) = (K \times I, K \times \dot{I})$ and define $f: A \to X$ by the given map $K \to X$ on one piece and a constant map on the other piece; then f can be extended over B because the obstructions lie in the trivial groups $\pi_i(X)$.

Second, as a particular case, a finite-dimensional complex K is contractible if (and only if) $\pi_i(K)$ is trivial for all $i \leq \dim K$.

Suppose f,g are two maps $X^n \to Y$ which agree on X^{n-1}. Then for each n-cell of B which is not in A we get a map $S^n \to Y$ by taking f and g on the two hemispheres. The resulting cochain of $C^n(B,A;\pi_n(Y))$ is called the *difference cochain* of f and g, denoted $d(f,g)$. The main facts about the difference cochain are the following (we omit the proofs).

Coboundary formula

$\delta d(f,g) = c(g) - c(f)$.

Addition formula

$d(f,g) + d(g,h) = d(f,h)$.

Any cochain can be realized as a difference cochain. This is more precisely stated in Proposition 1.

Proposition 1

Given $f\colon X^n \to Y$ as before, and given any cochain $d \in C^n(B,A;\pi_n(Y))$, there exists $g\colon X^n \to Y$ such that $d(f,g)$ is defined and is equal to d.

The proof is an easy consequence of the homotopy extension property of complexes.

Denote by $[c(f)]$ the cohomology class of $c(f)$.

Theorem 3

There is a map $g\colon X^{n+1} \to Y$ which agrees with f on X^{n-1} if and only if $[c(f)] = 0$, that is, $c(f)$ is a coboundary.

PROOF: Suppose $c(f) = \delta d$; choose g which agrees with f on X^{n-1} and such that $d(f,g) = -d$; then $c(f) = \delta d = -\delta d(f,g) = c(f) - c(g)$. Thus $c(g) = 0$, and so g extends over X^{n+1}. The reader can easily supply the proof of the converse.

We now want to interpret the above results in terms of extension of homotopies. Let K be a complex. Let $(B,A) = (K \times I, K \times \dot{I})$ and let $X^n = A \cup B^n$ as before. Now suppose that f,g are two maps $K \to Y$ which agree on K^{n-1}. Then there is an obvious map $a\colon X^n \to Y$. Moreover, $c(a)$ corresponds to $d(f,g)$—more precisely, to $d(f\,|\,K^n, g\,|\,K^n)$—under the identification of $C^{n+1}(B,A;\pi_n(Y))$ with $C^n(K;\pi_n(Y))$. The following theorem is merely a translation of the previous results into the present context.

Theorem 4

The restrictions of f and g to K^n are homotopic, rel K^{n-1}, if and only if $d(f,g) = 0 \in C^n(K;\pi_n(Y))$. They are homotopic, rel K^{n-2}, if and only if $[d(f,g)] = 0 \in H^n(K;\pi_n(Y))$.

APPLICATIONS OF OBSTRUCTION THEORY

In this section we will prove Theorem 1. But first we present a homotopy classification theorem. Let Y have a base point $*$; we will also denote by $*$ the corresponding constant map of any space into Y.

Theorem 5 (Hopf-Whitney)

Let K be a complex of dimension n, and let Y be $(n-1)$-connected. Then there is a one-to-one correspondence $k\colon [K,Y] \leftrightarrow H^n(K;\pi_n(Y))$.

PROOF: Denote by $[\![K,Y]\!]$ the homotopy classes, rel K^{n-2}, of maps $(K,K^{n-1}) \to (Y,*)$. Then there is a natural function $\theta\colon [\![K,Y]\!] \to [K,Y]$. We first prove that θ is 1–1 and onto.

Given any map $f: K \to Y$, the restriction $f \mid K^{n-1}$ is null-homotopic since Y is $(n-1)$-connected. By homotopy extension we can find a map $g: (K,K^{n-1}) \to (Y,*)$ such that $g \simeq f$. This proves that θ is onto.

To prove that θ is 1–1, let f,g be two maps $(K,K^{n-1}) \to (Y,*)$ such that $f \simeq g: K \to Y$. Let $h: K \times I \to Y$ be the homotopy. Write h' for the restriction of h to $K^{n-2} \times I$.

Consider $K^{n-2} \times I \times I$. We have $h': K^{n-2} \times I \times 0 \to Y$ and

$$*: K^{n-2} \times I \times 1 \to Y$$

Now $h(K^{n-1} \times \dot{I}) = *$; so $h': K^{n-2} \times \dot{I} \times 0 \to *$, and thus we have a partial homotopy between h' and $*: K^{n-2} \times I \to Y$, given by

$$H: K^{n-2} \times \dot{I} \times I \to *$$

Since Y is $(n-1)$-connected and $\dim (K^{n-2} \times I \times I) \leq n$, the obstructions to extending H over $K^{n-2} \times I \times I$ lie in groups with zero coefficients. Thus we have a map $H: K^{n-2} \times I \times I \to Y$, which is a homotopy between h' and $*$ such that $H: K^{n-2} \times \dot{I} \times I \to *$.

Now consider $K \times I \times I$. We have $h: K \times I \times 0 \to Y$ and

$$H: K^{n-2} \times I \times I \to Y$$

where H is a homotopy of $h \mid (K^{n-2} \times I)$. We can extend H to a homotopy $\bar{H}: K \times I \times I \to Y$, using (only) the homotopy extension theorem.

Now

$$\bar{H}(x,0,0) = h(x,0) = f(x)$$
$$\bar{H}(x,0,1) = f'(x)$$
$$\bar{H}(x,1,1) = g'(x)$$
$$\bar{H}(x,1,0) = h(x,1) = g(x)$$

where f' and g' are defined by the above equations. Moreover, $\bar{H} = *$ on $(K^{n-2} \times \dot{I} \times I) \cup (K^{n-2} \times I \times 1)$. Thus, $f \simeq f' \simeq g' \simeq g: K \to Y$, where all homotopies are rel K^{n-2}. This completes the proof that θ is 1–1.

We next define a transformation $k: [\![K, Y]\!] \to H^n(K;\pi_n(Y))$ and prove it 1–1 and onto.

Given $f: (K,K^{n-1}) \to (Y,*)$, let $k(f)$ be the cohomology class of $d(f,*)$—more precisely, $d(f \mid K^n,*)$—in $H^n(K;\pi_n(Y))$. This makes sense because $\dim K = n$ and hence any n-cochain is a cocycle.

Let f,g be two maps $(K,K^{n-1}) \to (Y,*)$. Then $k(f) - k(g) = [d(f,*) - d(g,*)] = [d(f,g)]$. But $[d(f,g)] = 0$ if and only if $f \simeq g$ rel K^{n-2}. This shows that $k: [\![K, Y]\!] \to H^n(K;\pi_n(Y))$ is well defined and 1–1.

Any cocycle $u \in Z^n(K;\pi_n(Y))$ may be realized as $d(f,*)$ for some map $f\colon (K,K^{n-1}) \to (Y,*)$. Then $k(f) = [d(f,*)] = [u]$, so that k is onto.

This proves the theorem. The correspondence $k\theta^{-1}$,

$$[K,Y] \xleftarrow[\approx]{\theta} [\![K,Y]\!] \xrightarrow[\approx]{k} H^n(K;\pi_n(Y))$$

may be described as follows: Every class $\alpha \in [K,Y]$ contains a map f such that $f \mid K^{n-1} = *$. Then $d(f,*)$ is defined and is a cocycle, and $k\theta^{-1}\colon \alpha \to [d(f,*)]$.

We are now ready to prove Theorem 1. If X is a space having the homotopy type of a CW-complex K, we may replace X by K without affecting the conclusion of the theorem. We therefore consider a CW-complex K (with no restriction on its dimension). To simplify notation, write Y for $K(\pi,n)$. We must show that $[f] \to f^*(\iota_n)$ gives a one-to-one correspondence, $[K,Y] \leftrightarrow H^n(K;\pi)$.

Let i be the inclusion of K^n into K and consider the following diagram:

$$\begin{array}{ccc} [K,Y] & \xrightarrow{i^\#} & [K^n,Y] \\ & & \approx \big\downarrow k\cdot\theta^{-1} \\ H^n(K;\pi) & \xrightarrow{i^*} & H^n(K^n;\pi) \end{array}$$

Now i^* is 1–1; moreover, $i^\#$ is 1–1 since, if $f \simeq g\colon K^n \to Y$, then also $f \simeq g\colon K \to Y$, since the obstructions to extending the homotopy lie in the zero groups $\pi_m(Y)$, $m > n$.

We claim that the image of $i^\#$ corresponds to the image of i^* under the correspondence $k \cdot \theta^{-1}$. In fact the following statements are easily seen to be equivalent, where $f\colon (K^n,K^{n-1}) \to (Y,*)$:

1. $[f]$ is in the image of $i^\#$
2. f extends over all of K
3. f extends over K^{n+1}
4. $c(f) = 0 \in C^{n+1}(K;\pi)$
5. $\delta d(f,*) = 0$
6. the cohomology class of $d(f,*)$ is in the image of i^*

The equivalence of (2) and (3), for example, follows from the fact that the higher obstructions to extending f lie in groups with zero coefficients $\pi_m(Y)$, $m > n$.

This proves that by going around the above diagram we get a correspondence $\varphi = (i^*)^{-1}(k \cdot \theta^{-1})(i^\#)\colon [K,Y] \to H^n(K;\pi)$, which is 1–1 and onto.

It is not immediately obvious that φ is the advertised correspondence $[f] \to f^*(\iota_n)$. One can verify that the correspondence $k \cdot \theta^{-1}$ is natural and hence that φ is natural, i.e., if A is another complex, any map $a: K \to A$ gives a commutative diagram as shown below.

$$\begin{array}{ccc} [K,Y] & \xleftarrow{a^\#} & [A,Y] \\ \downarrow{\scriptstyle\varphi} & & \downarrow{\scriptstyle\varphi} \\ H^n(K;\pi) & \xleftarrow{a^*} & H^n(A;\pi) \end{array}$$

Now read Y for A and f for a, and observe that φ takes the identity map id_Y of Y into ι_n. Thus $\varphi([f]) = \varphi(f^\#([id_Y])) = f^*(\varphi[id_Y]) = f^*(\iota_n)$.

This completes the proof of Theorem 1.

DISCUSSION

We have suggested that additional algebraic structure, namely, cohomology operations, will be a useful tool in the study of geometric problems. With the help of obstruction theory, the classification of cohomology operations has been reduced (Theorem 2) to the calculation of the cohomology of $K(\pi,n)$ spaces. We will pursue this calculation in some special cases in Chapter 9.

The $K(\pi,n)$ spaces were first studied in detail by Eilenberg and MacLane, and they are often referred to as "Eilenberg-MacLane spaces." Later work includes many calculations by Serre and Cartan.

Another possible approach is the construction of explicit operations. In the next chapter we construct an extremely important family of operations, due to Steenrod, who discovered them and proved many of their properties.

The cohomology operations considered in this chapter are sometimes called *primary* cohomology operations. In Chapter 16 we will introduce some more complicated constructions, all of which are known as higher order cohomology operations but are not cohomology operations in the sense of the present chapter.

In this chapter we have referred to the Hurewicz theorem, which will be proved in Chapter 8, and the cellular approximation theorem, which will be stated in Chapter 13. We have also appealed to the following theorem of J. H. C. Whitehead: A map of nice spaces inducing an isomorphism on each homotopy group is a homotopy equivalence.

REFERENCES

General

1. D. Husemoller [1].†
2. E. H. Spanier [1].
3. N. E. Steenrod [2].

K(π,*n*)

1. H. Cartan [5].
2. S. Eilenberg and S. MacLane [1,2].
3. J.-P. Serre [2].

CW-complexes and the Whitehead theorem

1. P. J. Hilton [1].
2. G. W. Whitehead [1].
3. J. H. C. Whitehead [1].

Obstruction theory

1. S. Eilenberg [1].
2. S.-T. Hu [1].
3. N. E. Steenrod [1].
4. G. W. Whitehead [1].

† Complete bibliographical material may be found in the Bibliography at the end of the book, where the authors' names are listed alphabetically.

CHAPTER 2

CONSTRUCTION OF THE STEENROD SQUARES

The object of this chapter is to construct the Steenrod squares. These are cohomology operations of type $(Z_2, n; Z_2, n+i)$.

We remark now, once and for all, that there are analogous operations for Z_p coefficients, where p is an odd prime; but they will not be treated in this book.

THE COMPLEX $K(Z_2,1)$

In Chapter 1 we constructed a CW-complex $K(\pi,n)$ for any abelian group π where $n \geq 2$. We now have need for an explicit complex $K(Z_2,1)$.

Proposition 1

Let $P = P(\infty) = \bigcup_n P(n)$ denote infinite-dimensional real projective space, the limit of $P(n)$ under the natural injections $P(n) \to P(n+1)$. Then P is a $K(Z_2,1)$ space.

PROOF: The n-sphere S^n is a covering space for $P(n)$ with covering group Z_2. From the exact homotopy sequence of this covering, it follows that $\pi_1(P(n)) = Z_2$ and that $\pi_i(P(n))$ is trivial for $1 < i < n$. The result follows.

Let S^∞ denote the infinite-dimensional sphere, i.e., the union of all S^n under the natural injections $S^n \to S^{n+1}$. We can easily give a cell structure for S^∞ as a CW-complex. In each dimension $i \geq 0$, we have two cells, which will be denoted d_i and Td_i. The action of the homology boundary ∂ is given by $\partial d_i = d_{i-1} + (-1)^i Td_{i-1}$, with $\partial T = T\partial$, $TT = 1$. (A sketch of S^2 will make this plausible.)

We can compute the homology of S^∞ from these formulas, verifying that S^∞ is acyclic. Indeed, in even dimensions the only non-zero cycles are generated by $d_{2j} - Td_{2j} = \partial d_{2j+1}$; in odd dimensions, only the sign is changed in the argument.

The homology of P is more interesting. We obtain a cell structure for $P = P(\infty)$ by collapsing S^∞ under the action of Z_2 considered as generated by T; in other words, identify d_i with Td_i for every $i \geq 0$. Thus P has exactly one cell in each dimension, denoted by e_i; and the boundary formula is $\partial e_{2j} = 2e_{2j-1}$, $\partial e_{2j-1} = 0$. Therefore $\tilde{H}_i(P; Z)$ must be Z_2 for i odd, 0 for i even.

By the universal coefficient theorems, since $H_*(P; Z)$ is Z_2 in odd dimensions, we have also the following: $H^*(P; Z)$ is Z_2 in positive even dimensions; $H_*(P; Z_2)$ and likewise $H^*(P; Z_2)$ are Z_2 in all dimensions.

Thus we know the homology and cohomology groups of $K(Z_2, 1)$ in all dimensions and for any coefficients and in particular for coefficients Z or Z_2. We turn to the calculation of the cohomology ring.

Let W denote the chain complex of S^∞. Then W is a Z_2-free acyclic chain complex with two generators in each dimension $i \geq 0$.

To compute the ring structure of $H^*(P)$, we must give a diagonal map for W. The action of T on W gives an action of T on $W \otimes W$ by $T(u \otimes v) = T(u) \otimes T(v)$.

Define $r: W \to W \otimes W$ by

$$r(d_i) = \textstyle\sum_{0 \leq j \leq i} (-1)^{j(i-j)} d_j \otimes T^j d_{i-j}$$
$$r(Td_i) = T(r(d_i))$$

where T^j of course denotes T or 1 according to whether j is odd or even. Then r is a chain map with respect to the usual boundary in $W \otimes W$, namely, $\partial(u \otimes v) = (\partial u) \otimes v + (-1)^{\deg u} u \otimes (\partial v)$. The verification is direct but tedious, and we omit it.

If h denotes the diagonal map of Z_2, then it is clear from the definitions that r is *h-equivariant*, that is, $r(gw) = h(g)r(w)$ for $g \in Z_2$ and $w \in W$. Therefore r induces a chain map $s: W/T \to W/T \otimes W/T$, where $W/T = W/Z_2$ is the chain complex of $P = P(\infty)$. Explicitly,

$$s(e_i) = \textstyle\sum_{0 \leq j \leq i} (-1)^{j(i-j)} e_j \otimes e_{i-j}$$

This map s is a chain approximation to the diagonal map Δ of P, and so we may use it to find cup products in $H^*(P; Z_2)$. Let α_i denote the nontrivial element of $H^i(P;Z_2)=Z_2$. The summation for $s(e_{j+k})$ contains a term $e_j \otimes e_k$ with coefficient 1 mod 2. Therefore,

$$\begin{aligned}\langle \alpha_j \cup \alpha_k, e_{j+k}\rangle &= \langle \Delta^*(\alpha_j \times \alpha_k), e_{j+k}\rangle \\ &= \langle \alpha_j \times \alpha_k, s(e_{j+k})\rangle \\ &\equiv 1 \pmod 2\end{aligned}$$

so that $\alpha_j \cup \alpha_k = \alpha_{j+k}$. We have indicated the proof of the following result.

Proposition 2

$H^*(Z_2,1;Z_2)=H^*(P;Z_2)$, as a ring, is the polynomial ring $Z_2[\alpha_1]$ on one generator, the non-zero one-dimensional class α_1.

THE ACYCLIC CARRIER THEOREM

To state the fundamental theorem on acyclic carriers, we need some terminology. Let π and G be groups (not necessarily abelian) and let $Z[\pi]$ denote the group ring of π. Let K be a π-free chain complex with a $Z[\pi]$-basis B of homogeneous elements, called "cells." For two cells $\sigma,\tau \in B$, let $[\tau:\sigma]$ denote the coefficient of σ in $\partial\tau$; this is an element in $Z[\pi]$. Let L be a chain complex on which G acts, and let h be a homomorphism $\pi \to G$.

Definition

An *h-equivariant carrier* $\mathcal{C}$ from K to L is a function $\mathcal{C}$ from B to the subcomplexes of L such that:

1. if $[\tau:\sigma]\neq 0$ then $\mathcal{C}\sigma \subset \mathcal{C}\tau$;
2. for $x \in \pi$ and $\sigma \in B$, $h(x)\mathcal{C}\sigma \subset \mathcal{C}\sigma$.

The carrier $\mathcal{C}$ is said to be *acyclic* if the subcomplex $\mathcal{C}\sigma$ is acyclic for every cell $\sigma \in B$. The h-chain map $f: K \to L$ is said to be *carried by* $\mathcal{C}$ if $f\sigma \in \sigma\mathcal{C}$ for every $\sigma \in B$.

Theorem 1

Let $\mathcal{C}$ be an acyclic carrier from K to L. Let K' be a subcomplex of K which is a $Z(\pi)$-free complex on a subset of B. Let $f: K' \to L$ be an h-equivariant chain map carried by $\mathcal{C}$. Then f extends over all of K to an h-equivariant chain map carried by $\mathcal{C}$. Moreover the extension is unique up to an h-equivariant chain homotopy carried by $\mathcal{C}$.

Note the important special case where K' is empty.

The proof proceeds by induction on the dimension; suppose that f has been extended over all of K^q and consider a $(q+1)$-cell $\tau \in B$. Then $\partial\tau = \sum a_i \sigma_i$ where $a_i = [\tau : \sigma_i] \in Z(\pi)$. Thus $f(\partial\tau) = \sum f(a_i\sigma_i) = \sum h(a_i) f(\sigma_i)$, which is in $\mathcal{C}\tau$ by properties (1) and (2). Since f is a chain map, $f(\partial\tau)$ is a cycle, but then, since $\mathcal{C}\tau$ is acyclic, there must exist x in $\mathcal{C}\sigma$ such that $\partial x = f(\partial\tau)$. Choose any such x, and put $f(\tau) = x$. This is the essential step in the construction; f is extended over K^{q+1} by requiring it to be h-invariant. Uniqueness is proved by applying the construction to the complex $K \times I$ and its subcomplex $K' \times I \cup K \times \dot{I}$.

CONSTRUCTION OF THE CUP-i PRODUCTS

Now let K be the chain complex of a simplicial complex, and let W be the Z_2-free complex discussed above. Define the action of Z_2 (generated by T) on $W \otimes K$ by $T(w \otimes k) = (Tw) \otimes k$, and on $K \otimes K$ by $T(x \otimes y) = (-1)^{dg\, x\, dg\, y}(y \otimes x)$. Define a carrier $\mathcal{C}$ from $W \otimes K$ to $K \otimes K$ by

$$\mathcal{C}: d_i \otimes \sigma \longrightarrow C(\sigma \times \sigma)$$

where by $C(\sigma \times \sigma)$ we mean the following: $K = C(X)$, the chain complex of the simplicial complex X. By the Eilenberg-Zilber theorem, there is a canonical chain-homotopy equivalence $\Psi: C(X \times X) \longrightarrow C(X) \otimes C(X)$. Then for σ a generator of K, i.e., a simplex of X, by $C(\sigma \times \sigma)$ we mean the subcomplex $\Psi C(\sigma \times \sigma)$ of $C(X) \otimes C(X)$. Then $\mathcal{C}$ is clearly acyclic and h-equivariant, where h is the identity map of Z_2. Therefore there exists an h-equivariant chain map

$$\varphi: W \otimes K \longrightarrow K \otimes K$$

carried by $\mathcal{C}$.

We should examine this φ because it plays a principal role in the definition of the squaring operations. Consider the restriction $\varphi_0 = \varphi \mid d_0 \otimes K$. This can be viewed as a map $K \longrightarrow K \otimes K$, and as such it is carried by the diagonal carrier. Thus it is a chain approximation to the diagonal, suitable for computing cup products in K. The same remarks apply to $T\varphi_0: \sigma \longrightarrow \varphi(Td_0 \otimes \sigma)$. Since both φ_0 and $T\varphi_0$ are carried by $\mathcal{C}$, they must be equivariantly homotopic. In fact it is not hard to verify that the chain homotopy is given by $\varphi_1: K \longrightarrow K \otimes K: \sigma \longrightarrow \varphi(d_1 \otimes \sigma)$. Further, φ_1 and $T\varphi_1$ are equivariantly homotopic homotopies; a homotopy is given by φ_2; and so forth.

We now use φ to construct cochain products.

Definition

For each integer $i \geq 0$, define a "cup-i product"

$$C^p(K) \otimes C^q(K) \to C^{p+q-i}(K)\colon (u,v) \to u \cup_i v$$

by the formula

$$(u \cup_i v)(c) = (u \otimes v)\varphi(d_i \otimes c) \qquad c \in C_{p+q-i}(K)$$

For the definition we must make an explicit choice of φ, but it will be seen that this choice is not essential.

Coboundary formula

$$\delta(u \cup_i v) = (-1)^i \, \delta u \cup_i v + (-1)^{i+p} u \cup_i \delta v - (-1)^i u \cup_{i-1} v - (-1)^{pq} v \cup_{i-1} u$$

(It is understood that $u \cup_{-1} v = 0$.)

The proof is a routine computation from the definitions: let c be a chain of $C_{p+q-i+1}(K)$; then

$$(\delta(u \cup_i v))(c) = (u \cup_i v)(\partial c) = (u \otimes v)\varphi(d_i \otimes \partial c)$$

By definition, $\partial(d_i \otimes c) = \partial d_i \otimes c + (-1)^i d_i \otimes \partial c$; so this becomes

$$\begin{aligned}
&(-1)^i (u \otimes v)\, \varphi\partial(d_i \otimes c) - (-1)^i (u \otimes v)\varphi(\partial d_i \otimes c) \\
&\quad = (-1)^i (u \otimes v)\, \partial\varphi(d_i \otimes c) - (-1)^i (u \otimes v)\varphi(d_{i-1} \otimes c) \\
&\qquad\qquad - (-1)^{2i} (u \otimes v)\varphi(Td_{i-1} \otimes c) \\
&\quad = (-1)^i \, \delta(u \otimes v)\varphi(d_i \otimes c) - (-1)^i (u \otimes v)\varphi(d_{i-1} \otimes c) \\
&\qquad\qquad - (u \otimes v) T\varphi(d_{i-1} \otimes c) \\
&\quad = (-1)^i \, \delta(u \otimes v)\varphi(d_i \otimes c) - (-1)^i (u \otimes v)\varphi(d_{i-1} \otimes c) \\
&\qquad\qquad - (-1)^{pq} (v \otimes u)\varphi(d_{i-1} \otimes c)
\end{aligned}$$

By definition of $\delta(u \otimes v)$, the first term may be written

$$(-1)^i (\delta u \otimes v)\varphi(d_i \otimes c) + (-1)^{i+p} (u \otimes \delta v)\varphi(d_i \otimes c)$$

and so we finally have, from the definitions,

$$\begin{aligned}
(\delta(u \cup_i v))(c) = (-1)^i \, \delta u \cup_i v + (-1)^{i+p} u \cup_i \delta v - (-1)^i u \cup_{i-1} v \\
- (-1)^{pq} v \cup_{i-1} u
\end{aligned}$$

which is the required formula.

THE SQUARING OPERATIONS

We emphasize that the cup-i products are defined on integral cochains and take values in integral cochains. But suppose $u \in C^p(K)$ is a cocycle mod 2, that is, $\delta u = 2a$, $a \in C^{p+1}(K)$. It follows from the coboundary formula that $u \cup_i u$ is also a cocycle mod 2. We can therefore define operations

$$Sq_i \colon Z^p(K;Z_2) \to Z^{2p-i}(K;Z_2) \colon u \to u \cup_i u$$

in the obvious way. Moreover, we can compose this with the natural projection of cocycles onto cohomology classes.

Lemma 1

The resulting function $Sq_i \colon Z^p(K;Z_2) \to H^{2p-i}(K;Z_2)$ is a homomorphism.

PROOF: Let c be a cochain of the appropriate dimension, and write down $Sq_i(u+v)(c)$. As expected, one obtains the terms $Sq_i(u)(c)$ and $Sq_i(v)(c)$ plus two cross terms, but the sum of the cross terms is a coboundary mod 2:

$$(u \cup_i v)(c) + (v \cup_i u)(c) = \delta(u \cup_{i+1} v)(c) \qquad \text{(mod 2)}$$

Lemma 2

If u is a coboundary, so also is $Sq_i(u)$.

PROOF: If $u = \delta a$, $Sq_i(u) = \delta(a \cup_i \delta a + a \cup_{i-1} a) \qquad \text{(mod 2)}$

Proposition 3

The above operation passes to a homomorphism:

$$Sq_i \colon H^p(K;Z_2) \to H^{2p-i}(K;Z_2)$$

This follows from the preceding lemmas. Of course "homomorphism" is understood in the sense of additive groups, not of rings.

Proposition 4

Let f be a continuous map $K \to L$. Then f^* commutes with Sq_i as in the diagram below.

$$\begin{array}{ccc} H^p(L;Z_2) & \xrightarrow{Sq_i} & H^{2p-i}(L;Z_2) \\ \downarrow{\scriptstyle f^*} & & \downarrow{\scriptstyle f^*} \\ H^p(K;Z_2) & \xrightarrow{Sq_i} & H^{2p-i}(K;Z_2) \end{array}$$

PROOF: By the simplicial approximation theorem, we may assume f is simplicial. Let u be a p-cochain of L. From the definitions, we have the formulas

$$f^*(Sq_i(u)) \colon c \to (u \otimes u)\varphi_L(d_i \otimes f(c)) = (u \otimes u)\varphi_L(1 \otimes f)(d_i \otimes c)$$
$$Sq_i(f^*(u)) \colon c \to (f^*u \otimes f^*u)\varphi_K(d_i \otimes c) = (u \otimes u)(f \otimes f)\varphi_K(d_i \otimes c)$$

where c is a $(2p-i)$-chain of K. But the two chain maps $\varphi_L(1 \otimes f)$ and $(f \otimes f)\varphi_K$ are both carried by the acyclic carrier $\mathcal{C}$ from $W \otimes K$ to $L \otimes L$

given by $\mathcal{C}(d_i \otimes \sigma) = C(f\sigma \times f\sigma)$. Thus they are equivariantly chain-homotopic, and hence the two images displayed above are cohomologous. In fact, if h is the homotopy, the difference between the above cochains is δg where $g(e) = (u \otimes u)h(d_i \otimes e)$.

The definition of the squaring operations is now complete, in the sense that we can draw the following inference.

Corollary 1

The operation Sq_i is independent of the choice of φ.

The corollary is proved by putting $K = L$ in Proposition 4, letting φ_K, φ_L denote two different choices of φ, and taking for f the identity map.

Proposition 5

If u is a cochain of dimension p, then $Sq_0(u)$ is the cup-product square u^2.

This follows from the remarks given after the definition of φ, since $(u \cup_0 u)(c) = (u \otimes u)\varphi_0(c)$ and φ_0 is suitable for computing cup products.

We have already begun to assemble the basic properties of the squaring operations, and henceforward it will be more convenient to modify the notation as follows.

Definition

Denote by Sq^i (with upper index) the natural homomorphisms

$$Sq^i\colon H^p(K;Z_2) \to H^{p+i}(K;Z_2) \qquad i = 0, 1, \ldots, p$$

given by $Sq^i = Sq_{p-i}$. For values of i outside the range $0 \leq i \leq p$, Sq^i is understood to be the zero homomorphism.

Thus Sq^i raises dimension by i in the cohomology of K.

COMPATIBILITY WITH COBOUNDARY AND SUSPENSION

We now wish to define squaring operations in relative cohomology. Let L be a subcomplex of K; we have an exact sequence, at the cochain level,

$$0 \to C^*(K,L) \xrightarrow{q^*} C^*(K) \xrightarrow{j^*} C^*(L) \to 0$$

We may assume that $\varphi_L = \varphi_K \mid W \otimes L$, since $\varphi_K(d_i \otimes \sigma) \in C(\sigma \times \sigma) \subset L \otimes L$ for $\sigma \in L$. This implies that for $u,v \in C^*(K)$, $j^*(u \cup_i v) = j^*u \cup_i j^*v$. Define relative cup-$i$ products as follows. Let $u,v \in C^*(K,L)$; then $j^*(q^*u \cup_i q^*v) = 0$, so that $(q^*u \cup_i q^*v)$ is in the image of q^*, by exactness; but q^* is one-to-one, and hence we may define $u \cup_i v$ as the unique cochain in $C^*(K,L)$ such that $q^*(u \cup_i v) = q^*u \cup_i q^*v$. It is trivial to verify that the coboundary formula for cup-i products carries over into this context. Therefore we obtain homomorphisms

$$Sq^i: H^p(K,L;Z_2) \to H^{p+i}(K,L;Z_2)$$

by the same process as in the absolute case. It is obvious from the definition that $q^*Sq^i = Sq^iq^*$.

In what follows, all coefficients are in Z_2 and we drop the Z_2 from the notation.

We recall the definition of the coboundary homomorphism $\delta^*: H^*(L) \to H^*(K,L)$. Let a be a cocycle and $\bar{a}$ its cohomology class, $a \in \bar{a} \in H^p(L)$. Then $a = j^*b$ for some $b \in C^p(K)$. Then $j^*(\delta b) = (\delta a) = 0$, and so $\delta b = q^*c$ for some $c \in C^{p+1}(K,L)$. Since q^* is one-to-one, c must be a cocycle, and by definition $\delta^*(a) = \bar{c}$, the cohomology class of c.

Proposition 6

Sq^i commutes with δ^* as in the diagram.

$$\begin{array}{ccc} H^p(L) & \xrightarrow{Sq^i} & H^{p+i}(L) \\ \downarrow{\scriptstyle \delta^*} & & \downarrow{\scriptstyle \delta^*} \\ H^{p+1}(K,L) & \xrightarrow{Sq^i} & H^{p+i+1}(K,L) \end{array} \qquad \text{(coefficients } Z_2\text{)}$$

PROOF: Using the notation of the last paragraph, $Sq^i\bar{a}$ and $Sq^i(\delta^*\bar{a})$ are represented by $a \cup_{p-i} a$ and $c \cup_{p+1-i} c$, respectively. Now

$$q^*(c \cup_{p+1-i} c) = q^*c \cup_{p+1-i} q^*c = \delta b \cup_{p+1-i} \delta b = \delta b' \qquad (\text{mod } 2)$$

where $b' = (b \cup_{p+1-i} \delta b) + (b \cup_{p-i} b)$. Moreover, $j^*(b') = 0 + (a \cup_{p-i} a)$. Therefore, by definition of δ^*,

$$\delta^*(Sq^i(\bar{a})) = \delta^*[a \cup_{p-i} a] = [c \cup_{p+1-i} c]$$

and the class on the right is $Sq^i(\delta^*(a))$.

Recall that, given any space X, we may define the *cone* over X and the *suspension* of X from the product space $X \times I$ by collapsing $X \times 0$ or $X \times \dot{I}$, respectively, to a point. In reduced cohomology, we have the *suspension isomorphism* $S^*: \tilde{H}^*(X) \to \tilde{H}^*(SX)$ defined by the composition

$$\tilde{H}^p(X) \underset{\approx}{\xrightarrow{\delta^*}} H^{p+1}(CX,X) \underset{\approx}{\longrightarrow} \tilde{H}^{p+1}(SX)$$

which raises dimension by one. The second isomorphism is proved by an excision argument and is based on a map. This, together with the naturality of the squaring operations and Proposition 6, yields the following fundamental property.

Proposition 7

The squares commute with suspension:

$$Sq^i \cdot S^* = S^* \cdot Sq^i \colon \tilde{H}^p(X) \to \tilde{H}^{p+i+1}(SX)$$

We can apply Proposition 7 to obtain an example of a non-trivial squaring operation which is not just a cup product. Let K denote the real projective plane; its cohomology ring with Z_2 coefficients is just the polynomial ring over Z_2 generated by the non-trivial one-dimensional cohomology class α and truncated by the relation $\alpha^3 = 0$. From Proposition 5, $Sq^1(\alpha) = Sq_0(\alpha) = \alpha^2$, so that $S^*Sq^1(\alpha)$ is non-zero. Hence $Sq^1S^*(\alpha)$ is also non-zero, showing that the operation Sq^1 is non-trivial in $H^2(SK; Z_2)$.

DISCUSSION

The squaring operations constructed in this chapter are a special case of the reduced power operations of Steenrod. These operations have been very important in the development of algebraic topology; most of this book is devoted to their properties and applications. And there are many more.

The specific construction we have given is neither the simplest possible nor the most subtle. Steenrod's original definition is more direct; his most recent definition is far more elegant. The construction we have given is adopted as a middle ground—one from which the algebraic properties are easily deduced, and yet the geometric genesis is not totally obliterated.

The reader will observe that we have defined the squaring operations in the simplicial cohomology theory, yet we will use these operations in singular theory. We justify this usage as follows. In their book (pp. 123–124), Steenrod and Epstein show that the squares, if defined for finite regular cell complexes, have unique extension to both singular and Čech theory for arbitrary pairs. But a finite regular cell complex has the homotopy type of a simplicial complex, on which we have defined the operations.

EXERCISE

1. Suppose the cocycle $u \in C^{2p}(X; Z)$ satisfies $\delta u = 2a$ for some a.
   i. Show that $u \cup_0 u + u \cup_1 u$ is a cocycle mod 4.
   ii. Define a natural operation, the *Pontrjagin square*,

$$P_2\colon H^{2p}(\ ; Z_2) \to H^{4p}(\ ; Z_4)$$

iii. Show that $\rho P_2(u) = u \cup u$, where $\rho: H^*(;Z_4) \to H^*(;Z_2)$ denotes reduction mod 2.

iv. Show that $P_2(u + v) = P_2(u) + P_2(v) + u \cup v$, where $u \cup v$ is computed with the non-trivial pairing $Z_2 \otimes Z_2 \to Z_4$.

REFERENCES

General

1. N. E. Steenrod [2].

Other definitions of the squares

1. N. E. Steenrod [3,5,6].
2. ——— and D. B. A. Epstein [1].

CHAPTER 3

PROPERTIES OF THE SQUARES

We assemble the fundamental properties of the squaring operations in an omnibus theorem.

Theorem 1

The operations Sq^i, defined (for $i \geq 0$) in the previous chapter, have the following properties:

0. Sq^i is a natural homomorphism $H^p(K,L; Z_2) \to H^{p+i}(K,L; Z_2)$
1. If $i > p$, $Sq^i(x) = 0$ for all $x \in H^p(K,L; Z_2)$
2. $Sq^i(x) = x^2$ for all $x \in H^i(K,L; Z_2)$
3. Sq^0 is the identity homomorphism
4. Sq^1 is the Bockstein homomorphism
5. $\delta^* Sq^i = Sq^i \delta^*$ where $\delta^*: H^*(L; Z_2) \to H^*(K,L; Z_2)$
6. *Cartan formula*. $Sq^i(xy) = \sum_j (Sq^j x)(Sq^{i-j} y)$
7. *Adem relations*: For $a < 2b$, $Sq^a Sq^b = \sum_c \binom{b-c-1}{a-2c} Sq^{a+b-c} Sq^c$ where the binomial coefficient is taken mod 2

We remark that the above properties completely characterize the squaring operations and may be taken as axioms, as is done in the book of Steenrod and Epstein.

Properties (0), (1), (2) and (5) have been proved in the last chapter. This chapter will be devoted to the proof of (3), (4), (6), and (7).

Sq^1 AND Sq^0

Let β denote the Bockstein homomorphism attached to the exact coefficient sequence $0 \to Z \to Z \to Z_2 \to 0$. Then β is a homomorphism $H^*(K,L; Z_2) \to H^*(K,L; Z)$, which raises dimension by one. It is defined on $x \in H^p(K,L; Z_2)$ as follows: represent the class x by a cocycle c; choose an integral cochain c' which maps to c under reduction mod 2; then $\delta c'$ is divisible by 2 and $\frac{1}{2}(\delta c')$ represents βx.

The composition of β and the reduction homomorphism gives a homomorphism

$$\delta_2: H^p(K,L; Z_2) \to H^{p+1}(K,L; Z_2)$$

which we also call "the Bockstein homomorphism"; in fact, it is the Bockstein of the sequence $0 \to Z_2 \to Z_4 \to Z_2 \to 0$. To show that this is Sq^1, we will use the following lemma, which in the light of (3) and (4) will be seen to be a special case of the Adem relations (7).

Lemma 1

$\delta_2 Sq^j = 0$ if j is odd; $\delta_2 Sq^j = Sq^{j+1}$ if j is even.

PROOF: Given $u \in H^p(K,L; Z_2)$, let c be an integral cochain such that the reduction mod 2 of c is in the class u. Then $Sq^j u$ is the class of $(c \cup_{p-j} c)$. Now $\delta c = 2a$ for some integral cochain $a \in C^{p+1}(K,L)$. Writing i for $(p-j)$, we have, by the coboundary formula,

$$\delta(c \cup_i c) = (-1)^i 2a \cup_i c + (-1)^j c \cup_i 2a - (-1)^i c \cup_{i-1} c - (-1)^p c \cup_{i-1} c$$

Thus $\delta_2(Sq^j u)$ is represented by the mod 2 cocycle

$$a \cup_i c + c \cup_i a + (s)(c \cup_{i-1} c)$$

where the coefficient s is 0 or 1 according to whether j is even or odd, respectively. But the sum of the first two terms is a coboundary, namely, $\delta(c \cup_{i+1} a)$ (mod 2), and the last term represents $(s)(Sq^{j+1}u)$. This proves the lemma.

As a special case of the lemma, $\delta_2 Sq^0 = Sq^1$. This shows that (4) follows from (3). It remains to prove (3).

Property (3) must be true in the real projective plane $P(2)$, for in that case $\delta_2 Sq^0(\alpha) = Sq^1(\alpha) = (\alpha^2) \neq 0$, and so $Sq^0(\alpha) \neq 0$, which proves $Sq^0(\alpha) = \alpha$ since α is the only non-zero element of $H^1(P(2); Z_2) = Z_2$. We can deduce that (3) is true in the circle S^1 by taking a map $f: S^1 \to P(2)$ such that $f^*: \alpha \to \sigma$, where σ denotes the generator of $H^*(S^1; Z_2)$, and applying the naturality condition: $Sq^0\sigma = Sq^0 f^*\alpha = f^* Sq^0 \alpha = f^*\alpha = \sigma$. Then (3) holds in every sphere S^n by suspension (Proposition 6 of Chapter 2).

Let K be a complex of dimension n; map K to S^n so that σ pulls back to a given class in $H^n(K;Z_2)$ (use the Hopf-Whitney theorem, Theorem 5 of Chapter 1). Then (3) holds for that class. Now let K be any complex, of unrestricted dimension; (3) must hold on any n-dimensional cohomology class in K because the injection j of the n-skeleton K^n into K induces a monomorphism $j^*: H^n(K;Z_2) \to H^n(K^n;Z_2)$ and the result follows as before by naturality. Thus (3) has been shown to hold in absolute cohomology.

Now let K,L be a pair and $K \cup CL$ the space obtained by attaching to K the cone over L, attached at the common subspace L. We then have isomorphisms, commuting with Sq^0,

$$H^*(K,L) \approx H^*(K \cup CL, CL) \approx \tilde{H}^*(K \cup CL)$$

The first isomorphism is by excision; the second, by contractibility of CL. This completes the proof of (3) and hence also of (4).

THE CARTAN FORMULA AND THE HOMOMORPHISM Sq

The Cartan formula (6) has two forms, one where we interpret the multiplication as the external cross product and another in which it is interpreted as the cup product. We will prove the first form and deduce the second as a corollary.

Consider the composition

$$W \otimes K \otimes L \xrightarrow{r \otimes 1} W \otimes W \otimes K \otimes L \xrightarrow{T} W \otimes K \otimes W \otimes L$$
$$\xrightarrow{\varphi_K \otimes \varphi_L} K \otimes K \otimes L \otimes L \xrightarrow{T} K \otimes L \otimes K \otimes L$$

where $r: W \to W \otimes W$ was defined in Chapter 2 and T permutes the second and third factors (we are not concerned with sign changes because we want conclusions in Z_2 coefficients). This composition, which we denote by $\varphi_{K \otimes L}$, is easily seen to be suitable for computations of cup-i products and hence Sq^i, in $K \otimes L$.

Using the same letters to denote (co-)homology classes or their representatives, and writing p,q,n for dim u, dim v, and $p+q-i$, respectively, we compute as follows:

$$\begin{aligned} Sq^i(u \times v)(a \otimes b) &= ((u \otimes v) \cup_n (u \otimes v))(a \otimes b) \\ &= (u \otimes v \otimes u \otimes v)\varphi_{K \otimes L}(d_n \otimes a \otimes b) \\ &= (u \otimes u \otimes v \otimes v)\sum \varphi_K(d_j \otimes a) \otimes T^j \varphi_L(d_{n-j} \otimes b) \end{aligned}$$

the last step using the definition of r, and the summation being over j, $0 \leq j \leq n$;

$$\begin{aligned} Sq^i(u \times v)(a \otimes b) &= \sum (u \cup_j u)(a) \otimes (v \cup_{n-j} v)(b) \\ &= \sum (Sq^{p-j}u)(a) \otimes (Sq^{q-n+j}v)(b) \\ &= \sum (Sq^{p-j}u \times Sq^{q-n+j}v)(a \otimes b) \end{aligned}$$

Hence, since $Sq^i x$ is zero for i outside the range $0 \leq i \leq \dim x$, we have

$$\begin{aligned} Sq^i(u \times v) &= \textstyle\sum_{j=0}^{n} Sq^{p-j}u \times Sq^{q-n+j}v \\ &= \textstyle\sum_{s=i-q}^{p} Sq^s u \times Sq^{i-s}v \qquad s = p - j \\ &= \textstyle\sum_{s=0}^{i} Sq^s u \times Sq^{i-s}v \end{aligned}$$

which completes the proof of the first form of (6).

In order to prove the cup-product form of (6), let Δ denote the diagonal map of K; if $x, y \in H^*(K; Z_2)$, $x \cup y = \Delta^*(x \times y)$, and so

$$\begin{aligned} Sq^i(x \cup y) &= Sq^i \Delta^*(x \times y) \\ &= \Delta^* Sq^i(x \times y) \\ &= \Delta^* \textstyle\sum_{j=0}^{i} Sq^j x \times Sq^{i-j} y \\ &= \textstyle\sum Sq^i x \cup Sq^{i-j} y \end{aligned}$$

This completes the proof of the Cartan formula (6) in both interpretations.

We remarked before that the Sq^i are homomorphisms only in the sense of groups; the Cartan formula makes it clear that they are not ring homomorphisms, but they can be combined into a ring homomorphism in the following sense.

Definition

Define $Sq\colon H^*(K; Z_2) \to H^*(K; Z_2)$ by

$$Sq(u) = \textstyle\sum_i Sq^i u$$

The sum is essentially finite; the image $Sq(u)$ is not in general homogeneous, i.e., it need not be contained in H^p for some p. (It is understood that each Sq^i is defined on nonhomogeneous elements $u \in H^*$ by requiring it to be additive.)

Proposition 1

Sq is a ring homomorphism.

PROOF: Clearly $Sq(u) \cup Sq(v) = (\sum_i Sq^i u) \cup (\sum_j Sq^j v)$ has $Sq^i(u \cup v)$ as its $(p+q+i)$-dimensional term, by the Cartan formula. Thus $Sq(u) \cup Sq(v) = Sq(u \cup v)$.

As an application of this homomorphism, we compute the Sq^i on any power of any one-dimensional cohomology class of any complex, as follows.

Proposition 2

For $u \in H^1(K;Z_2)$, $Sq^i(u^j) = \binom{j}{i} u^{j+i}$.

PROOF: $Sq(u) = Sq^0 u + Sq^1 u = u + u^2$ by properties (1) to (3). But Sq is a ring homomorphism. Therefore $Sq(u^j) = (u+u^2)^j = u^j \sum_k \binom{j}{k} u^k$, and the proposition follows by comparing coefficients.

In particular this gives us the action of all the Sq^i in $H^*(Z_2,1;Z_2)$, since this is generated as a ring by a one-dimensional class.

We pursue this line not only for its intrinsic interest but because it will serve us in the proof of the Adem relations.

SQUARES IN THE n-FOLD CARTESIAN PRODUCT OF $K(Z_2,1)$

Definition

Let K_n be the topological product of n copies of $K(Z_2,1)$. Here we may take for $K(Z_2,1)$ the complex $P(\infty)$ discussed in Chapter 2.

Since $H^*(Z_2,1;Z_2)$ is the polynomial ring on the one-dimensional class, it follows by the Künneth theorem that the ring $H^*(K_n;Z_2)$ is the polynomial ring over Z_2 on generators $x_1, \ldots, x_n$, where x_i is the non-trivial one-dimensional class of the ith copy of $K(Z_2,1)$. In this polynomial ring $Z_2[x_1,\ldots,x_n]$, we have the subring S of symmetric polynomials, which (by the fundamental theorem of symmetric algebra) may be written as $Z_2[\sigma_1,\ldots,\sigma_n]$ where σ_j is the elementary symmetric function of degree j (for example, $\sigma_1 = x_1 + \cdots + x_n$).

Proposition 3

In $H^*(K_n;Z_2)$, $Sq^i(\sigma_n) = \sigma_n \sigma_i$ $(1 \leq i \leq n)$.

PROOF: $Sq(\sigma_n) = Sq(\prod x_i) = \prod Sq(x_i) = \prod (x_i + x_i^2) = \sigma_n(\prod(1+x_i)) = \sigma_n \sum_{i=0}^n \sigma_i$. The result follows.

Corollary 1

In $H^*(Z_2,n;Z_2)$, $Sq^i \iota_n \neq 0$ for $0 \leq i \leq n$.

PROOF: By Theorem 1 of Chapter 1, we can find a map f of K_n into $K(Z_2,n)$ such that $f^*(\iota_n) = \sigma_n$. Then $f^* Sq^i(\iota_n) = Sq^i f^*(\iota_n) = Sq^i(\sigma_n) = \sigma_n \sigma_i \neq 0$, which proves the corollary.

We can show more; we can find a host of linearly independent elements in $H^*(Z_2,n;Z_2)$ by using compositions of the squares.

Notation

Given a sequence $I=\{i_1,\ldots,i_r\}$ of (strictly) positive integers, denote by Sq^I the composition $Sq^{i_1}\cdots Sq^{i_r}$. By convention, $Sq^I=Sq^0$, the identity, when I is the empty sequence (the unique sequence with $r=0$).

For example, $Sq^{\{2,1\}}(x)=Sq^2(Sq^1(x))$.

Definitions

A sequence I as above is *admissible* if $i_j \geq 2(i_{j+1})$ for every $j<r$. (This condition is vacuously satisfied if $r \leq 1$.) In this case we may also refer to Sq^I as admissible. The *length* of any sequence I is the number of terms, r in the above notation. The *degree* $d(I)$ is the sum of the terms, $\sum_j i_j$. (Thus Sq^I raises dimension by $d(I)$.) For an admissible sequence I, the *excess* $e(I)$ is $2i_1-d(I)$.

For the excess, we have

$$\begin{aligned} e(I) &= 2i_1 - d(I) \\ &= i_1 - i_2 - \cdots - i_r \\ &= (i_1-2i_2)+(i_2-2i_3)+\cdots+(i_r) \end{aligned}$$

The last expression justifies the name, but the first two are more convenient in practice.

Recall that S denotes the symmetric polynomial subring of the polynomial ring $H^*(K_n;Z_2)=Z_2[x_1,\ldots,x_n]$. We define an ordering on the monomials of S as follows: given any such monomial, write it as $m=\sigma_{j_1}^{e_1}\sigma_{j_2}^{e_2}\cdots\sigma_{j_s}^{e_s}$ with the j_k in decreasing order, $j_1>j_2>\cdots$. Then put $m<m'$ if $j_1<j_1'$ or if $j_1=j_1'$ and $(m/\sigma_{j_1})<(m'/\sigma_{j_1})$.

Theorem 2

If $d(I)\leq n$, then $Sq^I(\sigma_n)$ can be written $\sigma_n\cdot Q_I$ where $Q_I=\sigma_{i_1}\cdots\sigma_{i_r}+$ (a sum of monomials of lower order).

We prove this theorem by induction on the length of I; it reduces to the last proposition if $r=1$. Let r be the length of I, and write J for $i_2,\ldots,i_r$. We assume the result for sequences of length $<r$ (in particular for J); then, using the Cartan formula,

$$\begin{aligned} Sq^I(\sigma_n) &= Sq^{i_1}Sq^J(\sigma_n) = Sq^{i_1}(\sigma_n Q_J) \\ &= \textstyle\sum_{m=0}^{i_1} Sq^m(\sigma_n)Sq^{i_1-m}(Q_J) \\ &= \sigma_n\sigma_{i_1}Q_J + \textstyle\sum_{m=0}^{i_1-1}\sigma_n\sigma_m Sq^{i_1-m}(Q_J) \end{aligned}$$

By the induction hypothesis on Q_J, $Sq^I(\sigma_n)$ may be written

$$\sigma_n\sigma_{i_1}\sigma_{i_2}\cdots\sigma_{i_r} + \sigma_n\sigma_{i_1}\text{ (lower terms of } Q_J) + \textstyle\sum_{m=0}^{i_1-1} \sigma_n\sigma_m Sq^{i_1-m}(Q_J)$$

Observe that $Sq^j(\sigma_i) \leq \sigma_{i+j}$ and also that $Sq^i(\sigma_i) = \sigma_i^2 < \sigma_{2i}$. Therefore the largest possible term after the first term in the above expression is obtained from the last group of terms by taking $m = i_1 - i_2 + 1$, so that $i_1 - m = i_2 - 1$; thus this term is

$$x = \sigma_n\sigma_{i_1-i_2+1}\sigma_{2i_2-1}\sigma_{i_3}\cdots\sigma_{i_r}$$

Now I is admissible, and so $2i_2 - 1 < 2i_2 \leq i_1$, but this implies that $x < \sigma_n\sigma_{i_1}\sigma_{i_2}\cdots\sigma_{i_r}$, and the proof is complete.

As I runs through all admissible sequences of degree $\leq n$, the monomials $\sigma_I = \sigma_{i_1}\sigma_{i_2}\cdots\sigma_{i_r}$ are linearly independent in S and hence in $H^*(K_n; Z_2)$. From the above theorem, the $Sq^I(\sigma_n)$ are also linearly independent. We can draw the following inference.

Corollary 2

As I runs through the admissible sequences of degree $\leq n$, the elements $Sq^I(\iota_n)$ are linearly independent in $H^*(Z_2,n; Z_2)$.

Choose a map $f\colon K_n \to K(Z_2,n)$ such that $f^*(\iota_n) = \sigma_n$. Then the corollary follows from the preceding remarks.

Proposition 4

If $u \in H^n(K; Z_2)$ for any space K, and I has excess $e(I) > n$, then $Sq^I u = 0$. If $e(I) = n$, then $Sq^I u = (Sq^J u)^2$, where J denotes the sequence obtained from I by dropping i_1.

This proposition may be considered as a generalization of properties (1) and (2) of Theorem 1. To prove the first statement, note that if $e(I) = i_1 - i_2 - \cdots - i_r > n$, then $i_1 > n + i_2 + \cdots + i_r = \dim(Sq^J u)$, so that $Sq^I u = Sq^{i_1}(Sq^J u) = 0$ by (1). The second statement of the proposition follows in a similar way from (2).

These results are included in a theorem of Serre, which states that $H^*(Z_2,n; Z_2)$ is exactly the polynomial ring over Z_2 with generators $\{Sq^I(\iota_n)\}$, as I runs through all admissible sequences of excess less than n. We will establish Serre's theorem in Chapter 9, using spectral-sequence methods.

As a corollary to Serre's theorem, we mention that the map $f\colon K_n \to K(Z_2,n)$ such that $f^*(\iota_n) = \sigma_n$ clearly has the property that f^* is a monomorphism through dimension $2n$. We will use this fact in the proof of the Adem relations, and so we call attention to the fact that our proof of Theorem 1 will not be complete until we have proved Serre's theorem.

THE ADEM RELATIONS

We now discuss the Adem relations (7).

An Adem relation has the form

$$R = Sq^a Sq^b + \sum_{c=0}^{[a/2]} \binom{b-c-1}{a-2c} Sq^{a+b-c} Sq^c \equiv 0 \pmod 2$$

where $a < 2b$, and $[a/2]$ denotes the greatest integer $\leq a/2$. We usually drop the limits of summation from the expression, since the lower limit is implicit in the term Sq^c while the upper limit is implicit in the convention that the binomial coefficient $\binom{x}{y}$ is zero if $y < 0$. We also use the standard convention that $\binom{x}{y} = 0$ if $x < y$. (As an exercise in the use of these conventions, the reader may note that the Adem relations give $Sq^{2n-1} Sq^n = 0$ for every n.)

We will establish the Adem relations through a series of lemmas.

Lemma 2

Let y be a fixed cohomology class such that $R(y) = 0$ for every Adem relation R. Then $R(xy) = 0$ for every one-dimensional cohomology class x (and every R).

We will defer the proof of Lemma 2 to the end, since it is elementary but long and complicated.

Lemma 3

For every R and for every $n \geq 1$, $R(\sigma_n) = 0$ where $\sigma_n \in H^n(K_n; Z_2)$ as before.

PROOF: Let 1 denote the unit in the ring $H^*(K_n; Z_2)$; then $R(1) = 0$ for every R, by dimensional arguments. Then $R(x_1) = R(1x_1) = 0$ by Lemma 2; and finally $R(\sigma_n) = R(x_1 \cdots x_n) = 0$ by induction on n using Lemma 2.

Lemma 4

Let y be any cohomology class of dimension n of any space K, with Z_2 coefficients, and let $R = R(a,b)$ be the Adem relation for $Sq^a Sq^b$ where $a + b \leq n$. Then $R(y) = 0$.

PROOF: By Serre's theorem we have a map $f^*: H^*(Z_2,n; Z_2) \to H^*(X^n; Z_2)$ which takes ι_n to σ_n and is a monomorphism through dimension $2n$. We have $R(\sigma_n) = 0$ by Lemma 2, and so $R(\iota_n) = 0$, since these elements have dimension $n + a + b \leq 2n$. The result for y follows by naturality, using a map $g: K \to K(Z_2,n)$ such that $g^*(\iota_n) = y$.

Lemma 5

Let R be an Adem relation. If $R(y)=0$ for every class y of dimension p, then $R(z)=0$ for every class z of dimension $(p-1)$.

PROOF: Let u denote the generator of $H^1(S^1;Z_2)$. Clearly $Sq^i u=0$ for all $i>0$. Therefore, by the Cartan formula, $R(u\times z)=u\times R(z)$. But $u\times z$ has dimension p; hence $R(u\times z)=0$, and so $R(z)=0$.

The Adem relations follow easily from Lemma 4 and Lemma 5 by induction on dimension.

It remains to prove Lemma 2.

We begin by recalling the formula $\binom{p}{q}=\binom{p-1}{q-1}+\binom{p-1}{q}$, which holds for all p,q except for the case $p=q=0$.

Lemma 6

$\binom{p}{q}+\binom{p}{q+1}+\binom{p-1}{q-1}+\binom{p-1}{q+1}\equiv 0 \pmod 2$ except in the cases $(p=q=0)$ and $(p=0, q=-1)$.

This lemma follows from the formula just cited. (The easiest way to see the sense of these two formulas is to consider Pascal's triangle.)

To prove Lemma 2 is to show $R(xy)=0$ where x is any one-dimensional class and y has the property that $R(y)=0$ for every R. We begin by applying the Cartan formula to $Sq^b(xy)$; since $\dim x=1$, $Sq^b(xy)= xSq^b y+x^2Sq^{b-1}y$. Again by the Cartan formula,

$$\begin{aligned} Sq^aSq^b(xy) &= Sq^a(xSq^by+x^2Sq^{b-1}y)\\ &= xSq^aSq^by+x^2Sq^{a-1}Sq^by+x^2Sq^aSq^{b-1}y+0\\ &\quad +x^4Sq^{a-2}Sq^{b-1}y \end{aligned}$$

the zero coming from $Sq^1(x^2)$, which is zero mod 2. In a similar manner we find that

$$\begin{aligned} \sum (s)Sq^{a+b-c}Sq^c(xy) &= x\sum (s)Sq^{a+b-c}Sq^cy+x^2\sum (s)Sq^{a+b-c-1}Sq^cy\\ &\quad +x^2\sum (s)Sq^{a+b-c}Sq^{c-1}y+x^4\sum (s)Sq^{a+b-c-2}Sq^{c-1}y \end{aligned}$$

where $s=s(c)=\binom{b-c-1}{a-2c}$. In these two formulas, the first terms match, since

$$xSq^aSq^by+x\sum (s)Sq^{a+b-c}Sq^cy=xR(y)=0$$

Now $a<2b$ implies $(a-2)<2(b-1)$, and so the fourth terms also match: since $R(y)=0$ for all R, in particular, for $R(a-2, b-1)$,

$$\begin{aligned} Sq^{a-2}Sq^{b-1}y &= \textstyle\sum_c \binom{b-c-2}{a-2c-2}Sq^{a+b-c-3}Sq^cy\\ &= \textstyle\sum_{c'} \binom{b-c'-1}{a-2c'}Sq^{a+b-c'-2}Sq^{c'-1}y \end{aligned}$$

where $c' = c + 1$. Thus it remains to show that

$$Sq^a Sq^{b-1} y + \sum \binom{b-c-1}{a-2c-1} Sq^{a+b-c-1} Sq^c y$$
$$= \sum (s) Sq^{a+b-c-1} Sq^c y + \sum (s) Sq^{a+b-c} Sq^{c-1} y$$

where the second term on the left-hand side (LHS) replaces $Sq^{a-1}Sq^b y$, using $R(a-1, b)$. We consider three cases.

CASE 1: $a = 2b - 2$. Then $a - 2c = 2(b - c - 1)$, and so $(s) = \binom{k}{2k} = 0$ unless $k = 0$, that is, unless $c = b - 1$; so RHS $= Sq^a Sq^{b-1} y + Sq^{a+1} Sq^{b-2} y$. Similarly, $\binom{b-c-1}{a-2c-1} = \binom{k}{2k-1} = 0$ unless $k = 1$, that is, unless $c = b - 2$; so LHS $= Sq^a Sq^{b-1} y + Sq^{a+1} Sq^{b-2} y$, and the two sides are equal.

CASE 2: $a = 2b - 1$. Proved by a similar argument.

CASE 3: $a < 2b - 2$. Then, by $R(a, b - 1)$,

$$Sq^a Sq^{b-1} y = \sum_c \binom{b-c-2}{a-2c} Sq^{a+b-c-1} Sq^c y$$

Also,

$$\sum (s) Sq^{a+b-c} Sq^{c-1} y = \sum_c \binom{b-c-1}{a-2c} Sq^{a+b-c} Sq^{c-1} y$$
$$= \sum_{c'} \binom{b-c'-2}{a-2c'-2} Sq^{a+b-c'-1} Sq^{c'} y$$

where $c' = c - 1$, so we are reduced to verifying that

$$\binom{b-c-2}{a-2c} + \binom{b-c-1}{a-2c-1} \equiv \binom{b-c-1}{a-2c} + \binom{b-c-2}{a-2c-2} \pmod 2$$

But this follows from Lemma 5, with $p = b - c - 1$, $q = a - 2c - 1$. The exceptional cases are excluded automatically, because $p = 0$, $q = 0$ or -1 means $b = c + 1$, $a = 2c + 1$ or $2c$, respectively, contradicting the Case 3 hypothesis.

This completes the proof of Lemma 2.

We attach a short table of representative Adem relations.

$Sq^1Sq^1 = 0$, $Sq^1Sq^3 = 0, \ldots;$ $Sq^1Sq^{2n+1} = 0$

$Sq^1Sq^2 = Sq^3$, $Sq^1Sq^4 = Sq^5, \ldots;$ $Sq^1Sq^{2n} = Sq^{2n+1}$

$Sq^2Sq^2 = Sq^3Sq^1$, $Sq^2Sq^6 = Sq^7Sq^1, \ldots;$ $Sq^2Sq^{4n-2} = Sq^{4n-1}Sq^1$

$Sq^2Sq^3 = Sq^5 + Sq^4Sq^1, \ldots;$ $Sq^2Sq^{4n-1} = Sq^{4n+1} + Sq^{4n}Sq^1$

$Sq^2Sq^4 = Sq^6 + Sq^5Sq^1, \ldots;$ $Sq^2Sq^{4n} = Sq^{4n+2} + Sq^{4n+1}Sq^1$

$Sq^2Sq^5 = Sq^6Sq^1, \ldots;$ $Sq^2Sq^{4n+1} = Sq^{4n+2}Sq^1$

$Sq^3Sq^2 = 0, \ldots;$ $Sq^3Sq^{4n+2} = 0$

$Sq^3Sq^3 = Sq^5Sq^1; \ldots$

$Sq^{2n-1}Sq^n = 0$

DISCUSSION

Theorem 1 lists the major properties of the squaring operations, and, as we remarked, these properties characterize the squares uniquely. Parts (0) to (5) are due to Steenrod. The Cartan formula (6) was indeed discovered by Cartan; the Adem relations (7) were proved independently, and by very different methods, by Adem and Cartan.

Theorem 2, the surrounding material, and the proof we have given of the Adem relations include work of Cartan, Serre, and Thom. From our point of view this material will be amplified and completed by the calculations of Serre given in Chapter 9.

REFERENCES

General

1. N. E. Steenrod [2].
2. — and D. B. A. Epstein [1].

The Adem relations

1. J. Adem [1,2].
2. H. Cartan [3].

The Cartan formula

1. H. Cartan [5].

Related properties of the squares

1. H. Cartan [1,2,5].
2. J.-P. Serre [2].
3. N. E. Steenrod and D. B. A. Epstein [1].
4. R. Thom [1].
5. C. T. C. Wall [1].

CHAPTER 4

APPLICATION: THE HOPF INVARIANT

Let S^n denote an oriented n-sphere, where $n \geq 2$. Let there be given a map $f: S^{2n-1} \to S^n$. Consider S^{2n-1} as the boundary of an oriented $2n$-cell, and form the cell complex $K = S^n \cup_f e^{2n}$. Precisely, this is the complex formed from the disjoint union of S^n and e^{2n} by identifying each point in $S^{2n-1} = \dot{e}^{2n}$ with its image under f. Then the integral cohomology of K is zero except for the dimensions 0, n, and $2n$, and is Z in those three dimensions. Denote by σ and τ the generators determined by the given orientations of the cohomology groups in dimensions n and $2n$, respectively. Then the cup-product square σ^2 is some integral multiple of τ.

Definition

The *Hopf invariant* of f is the integer $H(f)$ such that $\sigma^2 = H(f) \cdot \tau$.

PROPERTIES OF THE HOPF INVARIANT

The homotopy type of K depends only on the homotopy class of the map f used to define K, and thus $H(f)$ also depends only on the homotopy class of f. We may therefore speak of the Hopf invariant of a homotopy class; and we have defined a transformation $H: \pi_{2n-1}(S^n) \to Z$, namely, that transformation which takes a homotopy class α into the integer $H(f)$ where f is any representative of α.

If n is odd, $H(f)=0$ for all f. This follows immediately from the anti-commutativity of the cup product: $\sigma^2=-\sigma^2$; hence $\sigma^2=0$.

If n is 2, 4, or 8, there exists a map $f\colon S^{2n-1}\to S^n$ with Hopf invariant one. For example, in the case $n=2$, f may be taken as the natural projection $f\colon S^3\to CP(1)=S^2$, viewing S^3 as the unit sphere in the complex plane C^2. Such f is the attaching map in the complex projective plane, $CP(2)=S^2\cup_f e^4$. Similarly, the cases $n=4$, $n=8$ correspond to the quaternionic plane $QP(2)=S^4\cup_f e^8$ and the Cayley plane $CayP(2)=S^8\cup_f e^{16}$, respectively. These maps are known as the *Hopf maps*.

For values of n other than 2, 4, or 8, it is now known that no map of Hopf invariant one can exist. This result is a very deep theorem and we do not prove it; Theorem 2 below is a partial result.

Proposition 1

If n is even, there exists a map $f\colon S^{2n-1}\to S^n$ with Hopf invariant two.

PROOF: We may consider the product space $S^n\times S^n$ as the cell complex formed by attaching a $2n$-cell to the wedge (one-point union) of two spheres $S^n\vee S^n$, using an attaching map $g\colon S^{2n-1}\to S^n\vee S^n$ as suggested by Figure 1. Let F denote the "folding map" $F\colon S^n\vee S^n\to S^n$, and let φ

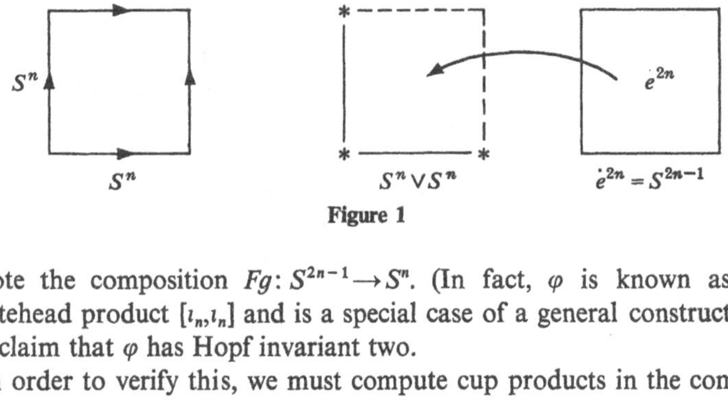

Figure 1

denote the composition $Fg\colon S^{2n-1}\to S^n$. (In fact, φ is known as the Whitehead product $[\iota_n,\iota_n]$ and is a special case of a general construction.) We claim that φ has Hopf invariant two.

In order to verify this, we must compute cup products in the complex $K=S^n\cup_\varphi e^{2n}$. Consider the diagram below, where the vertical maps are

$$\begin{array}{ccccc} e^{2n}\supset S^{2n-1} \xrightarrow{g} & S^n\vee S^n & \xrightarrow{\quad F \quad} & S^n & \\ & \downarrow & & \downarrow i & \\ S^n\times S^n = & (S^n\vee S^n)\cup_g e^{2n} & \xrightarrow{\bar F} & S^n\cup_\varphi e^{2n} & = K \end{array}$$

the inclusions. Now $\varphi=iFg\colon S^{2n-1}\to K$ is clearly null-homotopic, by definition of K. The map $\bar F$ is defined on $S^n\vee S^n$ by taking it equal to F there; but then it can be extended over all of $S^n\times S^n$ because $iFg=\varphi$ is null-homotopic. Moreover, $\bar F$ is a relative homeomorphism on the pair

(e^{2n},S^{2n-1}). Therefore $\bar{F}^*$ is an isomorphism in dimension $2n$. Now the folding map F has the property $F^*\sigma = \sigma_1 + \sigma_2$, where σ is the generator of $H^n(S^n)$ and σ_1,σ_2 are the generators of $H^n(S^n \vee S^n)$ corresponding to the two pieces. We denote generators of $H^{2n}(K)$ and $H^{2n}(S^n \times S^n)$ by τ and ρ, respectively. In the product $S^n \times S^n$, we have $\sigma_1^2 = \sigma_2^2 = 0$ and $\sigma_1\sigma_2 = \rho$ where we may choose ρ to make the sign positive. We also choose τ so that $\bar{F}^*(\tau) = \rho$. The Hopf invariant of φ is defined by $\sigma^2 = H(\varphi)\cdot\tau$. Then $\bar{F}^*\sigma^2 = H(\varphi)\bar{F}^*(\tau) = H(\varphi)\cdot\rho$. But

$$\begin{aligned}\bar{F}^*\sigma^2 = (\bar{F}^*\sigma)^2 = (F^*\sigma)^2 = (\sigma_1+\sigma_2)^2 &= \sigma_1^2+\sigma_2^2+\sigma_1\sigma_2+\sigma_2\sigma_1\\ &= 0+0+\rho+(-1)^{n^2}\rho\\ &= 2\rho\end{aligned}$$

so that $H(\varphi)\rho = 2\rho$ and $H(\varphi) = 2$, as was to be proved.

Proposition 2

The transformation $H\colon \pi_{2n-1}(S^n) \to Z$ is a homomorphism.

We first recall the definition of the additive group structure in $\pi_{2n-1}(S^n)$. We denote by Q the "pinching map"

$$Q\colon (e^{2n},S^{2n-1}) \to (e^{2n} \vee e^{2n},\, S^{2n-1} \vee S^{2n-1})$$

obtained by collapsing an equatorial $(2n-1)$-cell to a point. Let F denote the folding map $S^n \vee S^n \to S^n$, as before. Then if f and g represent elements of $\pi_{2n-1}(S^n)$, so that f and g are maps of S^{2n-1} into S^n, the composition

$$S^{2n-1} \xrightarrow{Q} S^{2n-1} \vee S^{2n-1} \xrightarrow{f\vee g} S^n \vee S^n \xrightarrow{F} S^n$$

represents, by definition, the sum of the homotopy classes of f and g.

To prove Proposition 2 we consider the following diagram, where p

$$\begin{array}{ccccc} = S^n \cup_{F(f\vee g)Q} e^{2n} & \xleftarrow{p} & L = (S^n \vee S^n)\cup_{(f\vee g)Q} e^{2n} & \xrightarrow{q} & M = (S^n \cup_f e^{2n}) \vee (S^n \cup_g e^{2n}) \\ \uparrow & & \uparrow & & \uparrow \\ S^n & \xleftarrow{F} & S^n \vee S^n & \xrightarrow{=} & S^n \vee S^n \end{array}$$

is constructed analogously to the map $\bar{F}$ of Proposition 1, q is based on the pinching of e^{2n} by Q, and the vertical maps are inclusions. Our notation for the generators in cohomology will be $\sigma \in H^n(K)$, $\tau \in H^{2n}(K)$, $\rho \in H^{2n}(L)$, τ_1 and τ_2 in $H^{2n}(M)$, and σ_1,σ_2 in $H^n(S^n \vee S^n)$. (We also use σ_1,σ_2 for the obvious classes in $H^n(L)$.)

By dcfinition, $\sigma^2 = H(f+g)\tau$. Thus, $p^*\sigma^2 = H(f+g)p^*\tau$. On the other hand, $p^*\sigma^2 = (\sigma_1+\sigma_2)^2$, as, in the proof of Proposition 1, for $\bar{F}$. Now in M, clearly $(\sigma_1+\sigma_2)^2 = H(f)\tau_1 + H(g)\tau_2$. Pulling this relation back to L, $(\sigma_1+\sigma_2)^2 = H(f)q^*\tau_1 + H(g)q^*\tau_2$. We can assume that ρ, τ_1, and τ_2 are chosen so that $p^*\tau = q^*\tau_1 = q^*\tau_2 = \rho$. Thus, in $H^*(L)$,

$$H(f+g)\rho = H(f+g)p^*\tau = p^*\sigma^2 = (\sigma_1 + \sigma_2)^2 = H(f)q^*\tau_1 + H(g)q^*\tau_2$$
$$= (H(f) + H(g))\rho$$

which implies $H(f+g) = H(f) + H(g)$, as was to be proved.

Corollary 1

If n is even, then $\pi_{2n-1}(S^n)$ contains Z (i.e., an infinite cyclic group) as a direct summand.

In fact, the cyclic group generated by the homotopy class of a map of Hopf invariant two must be mapped isomorphically onto the even integers by the homomorphism H.

DECOMPOSABLE OPERATIONS

We will be able to prove, by means of the squaring operations and the Adem relations, that there does not exist a map $f\colon S^{2n-1} \to S^n$ of Hopf invariant one except possibly when n is a power of 2. The clue to the method is the simple observation that f has an odd Hopf invariant, that is, σ^2 is an odd multiple of τ in $K = S^n \cup_f e^{2n}$, if and only if $Sq^n\sigma = \tau$ in the mod 2 cohomology of K. (Recall that $Sq^n\sigma = \sigma^2$.)

We will say that Sq^i is *decomposable* if $Sq^i = \sum_{t<i} a_t Sq^t$ where each a_t is a sequence of squaring operations, and we say Sq^i is indecomposable if no such relation exists. As examples, Sq^1 is obviously indecomposable; Sq^2 is indecomposable, since $Sq^1Sq^1 = 0$; Sq^3 is decomposable, since $Sq^3 = Sq^1Sq^2$ by the Adem relations. Less obviously, Sq^6 is decomposable by the Adem relation $Sq^2Sq^4 = Sq^6 + Sq^5Sq^1$.

Theorem 1

Sq^i is indecomposable if and only if i is a power of 2.

PROOF: We first suppose that i is a power of 2 and consider the generator α of $H^1(P(\infty); Z_2)$. Using the ring homomorphism Sq introduced in the last chapter, we have $Sq(\alpha^i) = (Sq\,\alpha)^i = (\alpha + \alpha^2)^i \equiv \alpha^i + \alpha^{2i}$ (mod 2). Thus $Sq^t(\alpha^i)$ is zero unless t is either 0 or i, while $Sq^i(\alpha^i) = \alpha^{2i}$. Now the fact that α^{2i} is non-zero shows that Sq^i is indecomposable; for otherwise $\alpha^{2i} = Sq^i(\alpha^i) = \sum_{t<i} a_t Sq^t(\alpha^i) = 0$.

To prove the converse, let $i = a + 2^k$ where $0 < a < 2^k$. Writing b for 2^k, the Adem relations give

$$Sq^aSq^b = \binom{b-1}{a} Sq^{a+b} + \sum_{c>0} \binom{b-c-1}{a-2c} Sq^{a+b-c}Sq^c$$

Since $b = 2^k$ is a power of 2, $\binom{b-1}{a} \equiv 1$ (mod 2), by the following lemma. Thus $Sq^i = Sq^{a+b}$ is decomposable.

Lemma 1

Let p be a prime, and let a and b have the p-adic expansions $a=\sum_{i=0}^{m} a_i p^i$, $b=\sum_{i=0}^{m} b_i p^i$ where $0 \leq a_i, b_i < p$. Then

$$\binom{b}{a} \equiv \prod_{i=0}^{m} \binom{b_i}{a_i} \qquad (\text{mod } p)$$

PROOF: In the polynomial ring $Z_p[x]$, $(1+x)^p = 1+x^p$. Thus $(1+x)^b = \prod (1+x)^{b_i p^i} \equiv \prod (1+x^{p^i})^{b_i}$. Now $\binom{b}{a}$ is the coefficient of x^a in this expansion, as is seen from the first expression; but the inspection of the last expression shows that this coefficient is precisely $\prod \binom{b_i}{a_i}$.

NON-EXISTENCE OF ELEMENTS OF HOPF INVARIANT ONE

We now state the main result of the chapter.

Theorem 2

If there exists a map $f\colon S^{2n-1} \rightarrow S^n$ of Hopf invariant one, then n is a power of 2.

We have already remarked that the existence of such a map f would imply that $Sq^n\sigma = \tau$ in the complex $K = S^n \cup_f e^{2n}$, where σ, τ are the generators of $H^*(K; Z_2)$ in dimensions n, $2n$, respectively. If n is not a power of 2, this is impossible, since then Sq^n is decomposable, whereas all $Sq^i\sigma$ must be zero in K (for dimensional reasons) if $0 < i < n$. This proves the theorem.

This result is closely related to the question of existence of multiplicative structures in Euclidean space R^n, as indicated by the following proposition.

Proposition 3

Suppose there is a map $\mu\colon R^n \times R^n \rightarrow R^n$ with a two-sided identity e, no zero-divisors, and the linearity property $(tx)y = t(xy) = x(ty)$, $t \in R$, where we have written xy for $\mu(x,y)$. Then $\pi_{2n-1}(S^n)$ contains an element of Hopf invariant one.

A proof is outlined in the exercises.

DISCUSSION

Our definition of the Hopf invariant is due to Steenrod; Hopf's original definition is more geometric but less manageable. In view of Proposition 3 and classical results of Hurwitz on the non-existence of norm-preserving

R-linear multiplications on R^n, much interest centered on the Hopf invariant question, that is, for what n is there a map from S^{2n-1} to S^n of Hopf invariant one?

Hopf's maps showed existence for $n = 2$, 4, or 8. The case of n odd was easily excluded. The case $n \equiv 2 \pmod 4$, $n > 2$, was excluded by G. W. Whitehead. Theorem 2 is due to Adem. The complete solution, namely, n must be 2, 4, or 8, is due to Adams.

The Whitehead product, discovered by J. H. C. Whitehead, is a homotopy operation in two variables, mapping $\pi_p(X) \otimes \pi_q(X) \to \pi_{p+q-1}(X)$. This operation is closely related to the Hopf invariant, as shown by G. W. Whitehead.

EXERCISES

1. Let $g: S^{n-1} \times S^{n-1} \to S^{n-1}$. Choose $y \in S^{n-1}$. Let α denote the degree of g restricted to $S^{n-1} \times y$ and β the degree of g restricted to $y \times S^{n-1}$. Show α and β are independent of the choice of y. Such a map g is said to be of *type* (α,β).
2. Let $g: S^{n-1} \times S^{n-1} \to S^{n-1}$. View S^{2n-1} as the join of S^{n-1} and S^{n-1}, with coordinates (a,t,b), $a,b \in S^{n-1}$, $t \in I$. View S^n as the suspension of S^{n-1}, with coordinates (c,t), $c \in S^{n-1}$, $t \in I$. Define $h(g): S^{2n-1} \to S^n$, the *Hopf construction* on g, by $h(g)(a,t,b) = (g(a,b), t)$. Prove $H(h(g)) = \alpha\beta$ if g has type (α,β).
3. Prove Proposition 3 by using the given map $\mu: R^n \times R^n \to R^n$ to construct a map $g: S^{n-1} \times S^{n-1} \to S^{n-1}$ of type (1,1).

REFERENCES

The Hopf invariant
1. H. Hopf [1,2].
2. N. E. Steenrod [4].
3. — and D. B. A. Epstein [1].

The Hopf invariant one problem
1. J. F. Adams [1,2].
2. J. Adem [1].
3. H. Cartan [6].

Whitehead products
1. P. J. Hilton [1].
2. G. W. Whitehead [2,4].

CHAPTER 5

APPLICATION: VECTOR FIELDS ON SPHERES

We consider the $(n-1)$-sphere S^{n-1} imbedded in R^n in the usual way. A *tangent vector field*, or simply a *vector field* on S^{n-1}, is a function assigning to each point of S^{n-1} a tangent vector at that point such that the vector thus defined varies continuously with the point. A *k-field* is an ordered set of k pointwise linearly independent tangent vector fields. We will be concerned with the problem of finding, for each n, the greatest k such that S^{n-1} admits a k-field; this maximum k will be denoted $K(n)$.

Clearly, for every n, $0 \leq K(n) \leq n-1$. We take as known the fact that when n is odd $K(n) = 0$; thus S^2, for example, does not admit a tangent vector field. (In fact an orientable smooth manifold admits a 1-field if and only if its Euler number vanishes.) For n even, $K(n) \geq 1$; it is easy to give a 1-field in this case: assign to the point $x = (x_1,\ldots,x_n)$ the vector $(x_2,-x_1,x_4,-x_3,\ldots,x_n,-x_{n-1})$. As in this example, we will identify a point in R^n with the vector from the origin to that point; and tangent vectors may be translated to the origin without explicit mention being made.

k-FIELDS AND $V_{n,k+1}$

Let $\{t_i(x)\}$ denote a k-field on S^{n-1}. By the Gram-Schmidt process we may assume that at each point x the vectors $\{t_i(x)\}$ form an orthonormal set. Then the set $\{t_1(x),\ldots,t_k(x),x\}$ is also orthonormal. Such a set is called a $(k+1)$-*frame*.

We denote by $V_{n,k+1}$ the set of $(k+1)$-frames in R^n. (The reader is warned that various authors use various subscripts $V_{i,j}$ for this set.)

Thus a k-field gives rise to a function $f\colon S^{n-1} \to V_{n,k+1}$. We also have a projection $p\colon V_{n,k+1} \to S^{n-1}$, taking $(v_1, \ldots, v_{k+1})$ into the last "coordinate" v_{k+1} considered as a point in S^{n-1}. Clearly the composition pf is the identity of S^{n-1}.

We will give the set $V_{n,k+1}$ a topology such that f and p are continuous. Then, if S^{n-1} admits a k-field, there exists a map $f\colon S^{n-1} \to V_{n,k+1}$ such that $pf\colon S^{n-1} \to S^{n-1}$ is the identity. Our line of attack to prove that S^{n-1} does *not* admit a k-field will be as follows. We will study the cohomology of the space $V_{n,k+1}$ and show that for certain values of (n,k), $H^{n-1}(V_{n,k+1};Z_2) = Z_2$, generated by $v = v_{n-1} = Sq^t z$ for some $z \in H^*(V_{n,k+1};Z_2)$ and some $t > 0$. Now suppose S^{n-1} admits a k-field, yielding $f\colon S^{n-1} \to V_{n,k+1}$. If we let σ denote the generator of $H^{n-1}(S^{n-1};Z_2)$, we must have $p^*\sigma = v$, $f^*v = \sigma$, since $pf = 1$. Now $\sigma = f^*v = f^*Sq^t z = Sq^t f^*z$; but f^*z is obviously zero because $\dim z < n-1$; and we thus arrive at a contradiction.

Our first task is therefore to give a topology for $V_{n,k+1}$. We will actually give a cell decomposition.

A CELL DECOMPOSITION OF $V_{n,k}$

Consider R^n as the space of column vectors with the usual basis. Let $\mathbf{O}(n)$ denote the orthogonal group, i.e., the group of orthogonal transformations of R^n. We view $\mathbf{O}(n)$ as a subgroup of the group of $n \times n$ matrices with determinant ± 1. $\mathbf{O}(n)$ acts on R^n from the left by matrix multiplication; it inherits a topology from R^{n^2}. We may consider $\mathbf{O}(n)$ as a subspace of $\mathbf{O}(n+1)$ by considering a given $n \times n$ matrix as an $(n+1) \times (n+1)$ matrix with zeroes in the last row and column except for a 1 on the main diagonal. Then $V_{n,k}$ may be identified with the left coset space $\mathbf{O}(n)/\mathbf{O}(n-k)$, by taking the last k columns of a representative matrix. This gives $V_{n,k}$ a topology. (In fact $V_{n,k}$ is a smooth manifold of dimension $\frac{1}{2}k(2n-k-1) = \sum_{i=n-k}^{n-1} i$, called a *Stiefel manifold*.) We have an obvious projection map p_{kj} when $0 \leq j \leq k$, $p_{kj}\colon V_{n,k} \to V_{n,j}$.

The projective space $P(n)$ may be considered as the space of lines through the origin in R^{n+1}. It will be momentarily confusing, but ultimately more convenient, to refer to this space as P_{n+1}. We write $P_{n,k} = P_n/P_{n-k}$, the identification space obtained by collapsing the subspace P_{n-k} to a point in P_n.

We have a map $\varphi: P_n \to \mathbf{O}(n)$ as follows: a point x in P_n is a line through the origin of R^n; $\varphi(x)$ will correspond to reflection through the hyperplane normal to the line x. If x is considered a point in R^n (and hence an n-vector), $\varphi(x)$ has the formula

$$y \to y - 2\frac{\langle x,y \rangle}{\|x\|^2}x$$

for $y \in R^n$. (Here $\langle\ ,\ \rangle$ denotes the scalar product and $\|\ \|$ the norm.)

Clearly $\varphi(x)$ is an orthogonal transformation. It is also evident that φ is 1–1. Actually φ is a homeomorphism into; we omit the verification. We have, for $m \leq n$, the diagram below, where the vertical maps are the stan-

$$\begin{array}{ccc} P_m & \xrightarrow{\varphi} & \mathbf{O}(m) \\ \downarrow & & \downarrow \\ P_n & \xrightarrow{\varphi} & \mathbf{O}(n) \end{array}$$

dard inclusions. In particular, taking $m = n - 1$, we obtain a map

$$\varphi: P_n/P_{n-1} \to \mathbf{O}(n)/\mathbf{O}(n-1)$$

that is, $P_{n,1} \to V_{n,1}$. Note that the slash represents, on the left, the identification of a subspace to a point; on the right, the formation of cosets.

Now $V_{n,1}$ is just the sphere S^{n-1}. On the other hand, P_n has a cell structure $P_n = e^0 \cup e^1 \cup \cdots \cup e^{n-2} \cup e^{n-1}$, so that $P_n = P_{n-1} \cup e^{n-1}$; thus $P_{n,1}$ is also S^{n-1}.

Proposition 1

$\varphi: P_{n,1} \to V_{n,1}$ is a homeomorphism.

Rather than give a detailed proof, we give a geometric description which should suggest the proof. Think of the case $n = 2$ so that $V_{n,1} = S^{n-1}$ is a circle, and we will consider $P_2 = P(1)$ as the lower semicircle, with its endpoints identified to a single point. Our transformation is based on the map $\varphi: P_n \to \mathbf{O}(n)$ which takes x to reflection in the hyperplane (line, when $n = 2$) normal to x, followed by projection onto $\mathbf{O}(n)/\mathbf{O}(n-1)$, i.e., considering the effect on the last coordinate. Using the language of a clock face for the case $n = 2$, we must take the unit vector u_2, that is, the unit vector in the 12 o'clock sense; choose x in P_2—a point between 3 and 9 o'clock; reflect u_2 in the line normal to the vector to x; and consider the image as a point on the circle $V_{2,1}$. The reader may verify that as x runs from 3 to 9 o'clock, the image of u_n runs from 12 to 12. For example, if x is at 4 o'clock, the normal line runs from 1 to 7 and u_2 reflects over to 2 o'clock.

Proposition 2

The matrix multiplication

$$\mu: (P_n \times V_{n-1,k-1}, P_{n-1} \times V_{n-1,k-1}) \to (V_{n,k}, V_{n-1,k-1})$$

is onto $V_{n,k}$ and is a relative homeomorphism.

This map is defined by identifying P_n as a subspace of $\mathbf{O}(n)$, using the map φ which is a homeomorphism into—and recall that $V_{n,k} = \mathbf{O}(n)/\mathbf{O}(n-k)$, etc.

We first prove that the inverse image of $V_{n-1,k-1}$ is precisely $P_{n-1} \times V_{n-1,k-1}$. Suppose $A,B \in P_n \times \mathbf{O}(n-1)$ are such that $AB\mathbf{O}(n-k)$ is contained in $\mathbf{O}(n-1)$; we must show that $A \in P_{n-1}$. Now the last vector is left fixed by B, by $\mathbf{O}(n-k)$, and by the product; hence certainly $A \in \mathbf{O}(n-1)$. But $P_n \cap \mathbf{O}(n-1) = P_{n-1}$, so that $A \in P_{n-1}$.

We prove next that μ is one-to-one on $(P_n - P_{n-1}) \times V_{n-1,k-1}$. Suppose $A,C \in P_n - P_{n-1}$, $B,D \in \mathbf{O}(n-1)$ are such that $AB\mathbf{O}(n-k) = CD\mathbf{O}(n-k)$. Then, since $B,D \in \mathbf{O}(n-1)$, we have $A\mathbf{O}(n-1) = C\mathbf{O}(n-1)$. Therefore $\varphi A = \varphi C$ in $V_{n,1}$; and so $A = C$, by Proposition 1. Hence $B\mathbf{O}(n-k) = D\mathbf{O}(n-k)$. But then B and D have the same image in $V_{n-1,k-1}$.

Finally we prove that μ is onto $V_{n,k}$. Suppose $A \in \mathbf{O}(n)$. By Proposition 1, $A\mathbf{O}(n-1) = C\mathbf{O}(n-1)$ for some $C \in P_n$. This means that for some $D \in \mathbf{O}(n-1)$, $A = CD$. But then $\mu(C,D)$ represents A.

We can now give the cell structure of $V_{n,k}$. Recall that $P_n = e^0 \cup e^1 \cup \cdots \cup e^{n-1}$. By a *normal cell* of $V_{n,k}$ we mean a cell of the form

$$e^{i_1-1} \times \cdots \times e^{i_r-1} \to P_{i_1} \times \cdots \times P_{i_r} \to V_{n,k} \qquad n \geq i_1 > \cdots > i_r > n-k$$

Theorem 1

The cells of $V_{n,k}$ are exactly the normal cells and the 0-cell corresponding to the identity matrix.

Consider the case $k = 1$; then we must have $i_1 = n$ and $r = 1$. Thus the only normal cell is $e^{n-1} \to P_n \to V_{n,1}$, and the theorem holds. We can therefore prove the theorem by induction on n and k, for $(n-k)$ fixed. Thus suppose the theorem is true for $V_{n-1,k-1}$; we are to prove it for $V_{n,k}$.

By Proposition 2, $V_{n,k} = V_{n-1,k-1} \cup_\mu ((P_n - P_{n-1}) \times V_{n-1,k-1})$. By the induction hypothesis, then,

$$V_{n,k} = (e^0 \cup \{e^{i_1-1} \times \cdots \times e^{i_r-1}\}) \cup_\mu (e^{n-1} \times (e^0 \cup \{e^{i_1-1} \times \cdots \times e^{i_r-1}\}))$$

where the cells in braces are subject to the restrictions $(n-1) \geq i_1 > \cdots > i_r > (n-k)$. But this agrees exactly with the assertion of the theorem for $V_{n,k}$. Thus the theorem is proved.

Corollary 1

If $2k \leq n$, then $P_{n,k}$ is the n-skeleton of $V_{n,k}$.

PROOF: When $2k \leq n$, the lowest dimension in which a normal cell may occur that has at least two factors is $(n-k)+(n-k+1)$, since it corresponds to the case $i_r = n-k+1$ and thus the two cells have dimension $i_r - 1 = n-k$ and $i_{r-1} - 1 \geq n-k+1$. But we have $(n-k)+(n-k+1) = 2n-2k+1 \geq n+1$. Thus the theorem above implies that the n-skeleton $(V_{n,k})^n$ consists of the 0-cell and of normal cells with only one factor:

$$(V_{n,k})^n = e^0 \cup e^{n-k} \cup e^{n-k+1} \cup \cdots \cup e^{n-1}$$

(As it happens, there is no n-cell.) Now $P_{n,k}$ is also a union of cells, of the same dimensions as in $(V_{n,k})^n$. Moreover, we have a map $\varphi\colon P_{n,k} \to V_{n,k}$, just as in the case $k=1$ discussed in Proposition 1; this map is a homeomorphism into, and its image is evidently $(V_{n,k})^n$. This proves the corollary.

THE COHOMOLOGY OF $P_{n,k}$

We turn now to consideration of the cohomology of these spaces. Recall that the ring $H^*(P_n; Z_2)$ is the truncated polynomial ring $Z_2[\alpha]/\alpha^n$. We have the sequence $P_{n-k} \xrightarrow{i} P_n \xrightarrow{j} P_{n,k} = P_n/P_{n-k}$, where i is the injection and j the identification map. Now $H^*(P_{n-k}; Z_2) = Z_2[\beta]/\beta^{n-k}$ and $\beta = i^*\alpha$. By the exact cohomology sequence of this pair, $\tilde{H}^q(P_{n,k}; Z_2) = Z_2$ in the range $(n-k) \leq q < n$ and is zero otherwise.

Let v_q denote the generator of $H^q(P_{n,k}; Z_2)$; then $j^*(v_q) = \alpha^q$. But $Sq^t(\alpha^q) = \binom{q}{t}\alpha^{q+t}$, and so $Sq^t v_q = \binom{q}{t} v_{q+t}$, provided $(n-k) \leq q$, $(q+t) < n$. In particular, $Sq^{k-1} v_{n-k} = \binom{n-k}{k-1} v_{n-1}$.

Now if $2k \leq n$, we have $(V_{n,k})^n = P_{n,k}$, and thus the cohomology of $V_{n,k}$ agrees with that of $P_{n,k}$ through dimension $n-1$. Therefore in this case $H^{n-1}(V_{n,k}; Z_2)$ is generated by v_{n-1} and $Sq^{k-1} v_{n-k} = \binom{n-k}{k-1} v_{n-1}$.

We now prove a non-existence theorem for tangent vector fields. Write the integer n in the form $2^m(2s+1)$; thus 2^m is the highest power of 2 dividing n.

Theorem 2

S^{n-1} does not admit a 2^m-field.

If $s=0$, then $2^m = n > n - 1$ and the result is trivial; if $m=0$, then n is odd and the result is known. We therefore suppose $m,s \geq 1$.

Recall the argument, given early in the chapter, that if S^{n-1} admits a 2^m-field, then there is a map $f\colon S^{n-1} \to V_{n,2^m+1}$ such that f followed by the projection $p: V_{n,2^m+1} \to S^{n-1}$ is the identity of S^{n-1}. Now $n = 2^m(2s+1) = 2(2^m)s + 2(2^{m-1}) \geq 2(2^m+1)$ since $m,s \geq 1$. Therefore, if we take $k = 2^m + 1$, we have $2k \leq n$, and the above results on the cohomology of $V_{n,k}$ are applicable: $H^{n-1}(V_{n,k}) = Z_2$ generated by v_{n-1}, and, since pf is the identity, $p^*(\sigma) = v_{n-1}$ where σ is the generator of $H^{n-1}(S^{n-1}; Z_2)$.

We have $Sq^{k-1}v_{n-k} = \binom{n-k}{k-1}v_{n-1}$; but $n - k = n - (2^m+1) = 2^{m+1}s - 1$ and $k - 1 = 2^m$; the binomial coefficient is seen to be non-zero by Lemma 2 of Chapter 4.

Thus we are led, from the assumption that a 2^m-field exists, to the contradiction

$$\sigma = f^*v_{n-1} = f^*(Sq^{2^m}v_{n-2^m-1}) = Sq^{2^m}(f^*v_{n-2^m-1}) = Sq^{2^m}(0) = 0$$

This completes the proof.

DISCUSSION

Write n in the form $n = 2^{4a+b}(2s+1)$ with $0 \leq b < 4$. Classical algebraic results of Hurwitz and Radon, treated recently from a modern viewpoint by Atiyah, Bott, and Shapiro, show that $K(n) \geq 2^b + 8a - 1$; that is, S^{n-1} admits a k-field with $k = 2^b + 8a - 1$.

On the other hand, Theorem 2, due to Steenrod and J. H. C. Whitehead, gives an upper bound for $K(n)$. For $n < 16$ and many other values, this upper bound coincides with the Hurwitz-Radon lower bound and completely determines $K(n)$.

The complete solution of this problem was given by Adams, who proved that for all n, the Hurwitz-Radon number is an upper bound as well; thus $K(n) = 2^b + 8a - 1$ for all n.

REFERENCES

General

1. J. Milnor [1].
2. N. E. Steenrod [1].
3. — and D. B. A. Epstein [1].

Cell structure of Stiefel manifolds

1. C. E. Miller [1].

The Hurwitz-Radon theorem

1. M. F. Atiyah, R. Bott, and A. Shapiro [1].
2. B. Eckmann [1].

Vector fields on spheres

1. J. F. Adams [3].
2. N. E. Steenrod and J. H. C. Whitehead [1].

CHAPTER 6

THE STEENROD ALGEBRA

In this chapter we describe the algebraic system formed by the Steenrod squaring operations and derive some of its properties.

GRADED MODULES AND ALGEBRAS

Let R be a commutative ring with unit.

By a *graded R-module M* we mean a sequence M_i $(i \geq 0)$ of unitary R-modules. A homomorphism f: $M \to N$ of graded R-modules is a sequence f_i of R-homomorphisms, f_i: $M_i \to N_i$. The tensor product $M \otimes N$ of two graded R-modules is defined by setting $(M \otimes N)_t = \sum_i M_i \otimes N_{t-i}$.

By a *graded R-algebra A* we mean a graded R-module with a multiplication φ: $A \otimes A \to A$, where φ is a homomorphism of graded R-modules and has a two sided unit. By a homomorphism of graded R-algebras we mean a homomorphism of graded R-modules respecting the multiplications and units.

A graded R-algebra A is *associative* if $\varphi \circ (\varphi \otimes 1) = \varphi \circ (1 \otimes \varphi)$, where 1 denotes the identity homomorphism $A \to A$; in other words, if the diagram below commutes.

$$\begin{array}{ccc} A \otimes A \otimes A & \xrightarrow{\varphi \otimes 1} & A \otimes A \\ {\scriptstyle 1 \otimes \varphi}\downarrow & & \downarrow{\scriptstyle \varphi} \\ A \otimes A & \xrightarrow{\varphi} & A \end{array}$$

We say that A is *commutative* if $\varphi \circ T = \varphi: A \otimes A \to A$, where $T: M \otimes N \to N \otimes M$ is defined by $T(m \otimes n) = (-1)^{ij}(n \otimes m)$, $i = \deg n$, $j = \deg m$.

If we are given an algebra homomorphism $\varepsilon: A \to R$, where R is considered as a graded R-algebra by the convention $R_0 = R$, $R_i = 0$ $(i \neq 0)$, then A is said to be *augmented*. An augmented R-algebra is *connected* if $\varepsilon: A_0 \to R$ is isomorphic.

If A and B are graded R-algebras, then $A \otimes B$ (the tensor product as graded R-modules) is given an algebra structure by defining $\varphi_{A \otimes B} = (\varphi_A \otimes \varphi_B) \circ (1 \otimes T \otimes 1)$. In other words, if we write products by juxtaposition, $(a_1 \otimes b_1)(a_2 \otimes b_2) = (-1)^k (a_1 a_2) \otimes (b_1 b_2)$, $k = (\deg a_2)(\deg b_1)$.

For example, let M be an R-module. We define the *tensor algebra* $\Gamma(M)$ as follows. Write M^0 for R, M^1 for M, M^2 for $M \otimes M$, and in general M^r for the r-fold tensor product $M \otimes \cdots \otimes M$. Then $\Gamma(M)$ is the graded R-algebra defined by $\Gamma(M)_r = M^r$ where the product is given by the usual isomorphism $M^s \otimes M^t \approx M^{s+t}$. The tensor algebra $\Gamma(M)$ is clearly associative, but not commutative.

THE STEENROD ALGEBRA $\mathcal{A}$

Now take $R = Z_2$, and let M be the graded Z_2-module such that $M_i = Z_2$ generated by the symbol Sq^i for each $i \geq 0$. $\Gamma(M)$ is thus bigraded. For each pair of integers (a,b) such that $0 < a < 2b$, let

$$R(a,b) = Sq^a \otimes Sq^b + \textstyle\sum_c \binom{b-c-1}{a-2c} Sq^{a+b-c} \otimes Sq^c$$

Let Q denote the ideal of $\Gamma(M)$ generated by all such $R(a,b)$ and $1 + Sq^0$.

Definition

The *Steenrod algebra* $\mathcal{A}$ is the quotient algebra $\Gamma(M)/Q$.

Note that $\mathcal{A}$ inherits a grading from the gradation on M. The elements of the Steenrod algebra are the polynomials in the Sq^i, $i \geq 0$ $(Sq^0 = 1)$, coefficients in Z_2, and subject to the Adem relations.

Theorem 1

The monomials Sq^I, as I runs through all admissible sequences, form a basis for $\mathcal{A}$ as a Z_2-module.

The fact that the Sq^I are linearly independent (for I admissible) follows from the linear independence of the elements $Sq^I(\iota_n) \in H^*(Z_2,n;\ Z_2)$ for all admissible I of degree $\leq n$, which is the assertion of Corollary 2 of Chapter 3.

The other part of the proof uses the Adem relations and a reduction argument. For any sequence $I = \{i_1, \ldots, i_r\}$ (not necessarily admissible), define the *moment* $m(I)$ by the formula $m(I) = \sum_s i_s s$. It is not hard to see that if I is not admissible, starting at the right and applying the Adem relation to the first pair i_s, i_{s+1} with $i_s < 2i_{s+1}$ leads to a sum of monomials of strictly lower moment than I. Since the moment function is bounded below, the process terminates and the admissible Sq^I actually span $\mathcal{A}$.

As an example, $\mathcal{A}_7$ has as basis Sq^7, $Sq^6 Sq^1$, $Sq^5 Sq^2$, $Sq^4 Sq^2 Sq^1$.

The basis given by Theorem 1 is known as the Serre-Cartan basis for $\mathcal{A}$. We will give a quite different basis later in this chapter.

DECOMPOSABLE ELEMENTS

Let A be a connected graded R-algebra, and let $\bar{A}$ denote the kernel of the augmentation ε; thus $\bar{A}$ is the ideal of elements of strictly positive degree. It is called the "augmentation ideal."

Definition

The ideal of *decomposable* elements of A is the ideal $\varphi(\bar{A} \otimes \bar{A}) \subset A$. The "set of indecomposable elements" of A is the R-module $Q(A) = \bar{A}/\varphi(\bar{A} \otimes \bar{A})$.

The language is somewhat misleading, since elements of $Q(A)$ are not elements of A but rather cosets. Note that $Q(A)$ is not an algebra, since it fails to have a unit.

The present definition of decomposable elements is consistent with the ad hoc definition given in Chapter 4 for Sq^i. There it was proved that Sq^i is decomposable if and only if i is not a power of 2. The following result is now obvious.

Corollary 1

$\{Sq^{2^i}\}$, $i \geq 0$, generate $\mathcal{A}$ as an algebra.

We remark that these elements do not generate $\mathcal{A}$ freely; for example, $Sq^2 Sq^2 = Sq^3 Sq^1 = Sq^1 Sq^2 Sq^1$ and $Sq^1 Sq^1 = 0$.

THE DIAGONAL MAP OF $\mathcal{A}$

Our next objective is to show that the algebra $\mathcal{A}$ possesses additional structure, namely, an algebra homomorphism $\mathcal{A} \rightarrow \mathcal{A} \otimes \mathcal{A}$. This mapping will prove to be a powerful tool in the study of $\mathcal{A}$.

As before, let M be the graded Z_2-module generated in each $i \geq 0$ by Sq^i. Define ψ: $\Gamma(M) \to \Gamma(M) \otimes \Gamma(M)$ by the formula

$$\psi(Sq^i) = \textstyle\sum_j Sq^j \otimes Sq^{i-j}$$

and the requirement that ψ be an algebra homomorphism. Thus

$$\psi(Sq^r \otimes Sq^s) = \psi(Sq^r) \otimes \psi(Sq^s) = (\textstyle\sum_a Sq^a \otimes Sq^{r-a}) \otimes (\textstyle\sum_b Sq^b \otimes Sq^{r-b}).$$

Theorem 2

The above ψ induces an algebra homomorphism ψ: $\mathcal{A} \to \mathcal{A} \otimes \mathcal{A}$.

PROOF: Let p: $\Gamma(M) \to \mathcal{A}$ be the natural projection. We must show that the kernel of p is contained in the kernel of the above ψ. As before, denote by K_n the n-fold Cartesian product of the space $P(\infty) = K(Z_2,1)$. Define a mapping w: $\mathcal{A} \to H^*(K_n;Z_2)$, raising degree by n, by $w(\theta) = \theta(\sigma_n)$; that is, w is evaluation on σ_n. By Theorem 2 of Chapter 3 and Theorem 1 above, we know that w is monomorphic through degree n (degree in $\mathcal{A}$). Let w' be defined similarly by evaluation on σ_{2n}, and consider the following diagram,

$$\begin{array}{ccccc}
\Gamma(M) & \xrightarrow{\;p\;} & \mathcal{A} & & \\
\downarrow{\scriptstyle\psi} & & \downarrow{\scriptstyle w \times w} & \searrow^{w'} & \\
\mathcal{A} \otimes \mathcal{A} & \xrightarrow{w \otimes w} & H^*(K_n) \otimes H^*(K_n) & \xrightarrow[\approx]{\alpha} & H^*(K_n \times K_n) = H^*(K_{2n})
\end{array}$$

where cohomology is understood to be with Z_2 coefficients, and the isomorphism α follows from the Künneth theorem (Z_2 is a field). We wish to show that the quadrilateral in this diagram is commutative.

Using the Cartan formula, it is clear that the diagram commutes on the generators Sq^i, as in the following computation:

$$\begin{aligned}
w'p(Sq^i) = w'(Sq^i) = Sq^i(\sigma_{2n}) &= Sq^i(\sigma_n \times \sigma_n) \\
&= \textstyle\sum_j Sq^j(\sigma_n) \times Sq^{i-j}(\sigma_n) \\
&= (w \otimes w)\psi(Sq^i)
\end{aligned}$$

Now p, ψ, and α are algebra homomorphisms, but w' is not; so commutativity in general does not follow immediately from commutativity on Sq^i. However, we may argue as follows. A basis for $M^r = M \otimes \cdots \otimes M$ is given by $\{Sq^{i_1} \otimes \cdots \otimes Sq^{i_r}\}$, where $I = \{i_1, \ldots, i_r\}$ need not be admissible. Abbreviating such elements by $Sq^I_\otimes$, we have

$$\begin{aligned}
w'p(Sq^I_\otimes) = w'(Sq^I) = Sq^I(\sigma_{2n}) &= Sq^{i_1} \cdots Sq^{i_{r-1}} \textstyle\sum_s Sq^s(\sigma_n) \times Sq^{i_r - s}(\sigma_n) \\
\cdots &= \textstyle\sum_{(I_1 + I_2 = I)} Sq^{I_1}(\sigma_n) \times Sq^{I_2}(\sigma_n)
\end{aligned}$$

Inspection shows that $\alpha(w \otimes w)\psi(Sq^I_\otimes)$ has the same formula, and this completes the proof that the diagram is commutative.

Now suppose $p(z)=0$, $z \in \Gamma(M)$. Then $0 = w'p(z) = \alpha(w \otimes w)\psi(z)$. But α is isomorphic, and $(w \otimes w)$ is monomorphic in a certain range; thus, for any z, we choose n sufficiently large (with respect to deg z) so that $(w \otimes w)$ is monomorphic when restricted to elements of degree $=$ deg z. Thus $\psi(z)=0$ and $(\ker p) \subset (\ker \psi)$. This completes the proof of the theorem.

HOPF ALGEBRAS

Let A be a connected graded R-algebra (with unit); the existence of the unit can be expressed by the fact that, in the following diagram, both compositions $A \to A$ are the identity map, where η, the "co-augmentation," is the inverse of the isomorphism $\varepsilon \mid A_0 : A_0 \to R$.

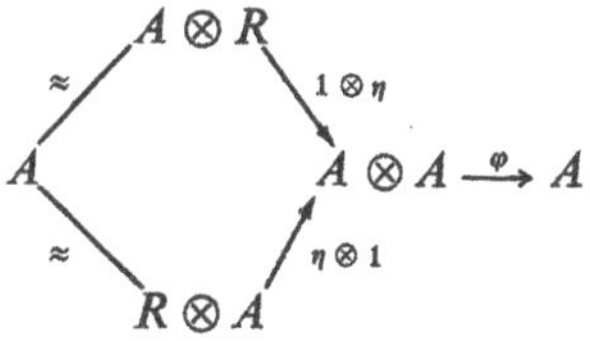

Now let A be a graded R-module with a given R-homomorphism $\varepsilon: A \to R$. We say that A is a *co-algebra* (with co-unit) if we are given an R-homomorphism $\psi: A \to A \otimes A$ such that both compositions are the identity in the dual diagram:

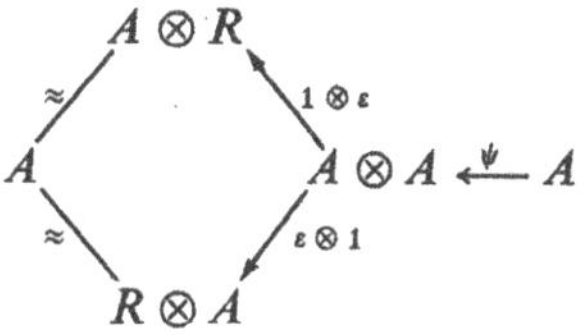

ψ is called the "co-multiplication" or "diagonal map."

Now let A be a connected graded R-algebra with augmentation ε. Suppose also that A has a co-algebra structure with co-unit ε, and that the diagonal map $\psi: A \to A \otimes A$ is a homomorphism of R-algebras. Then we say A is a (*connected*) *Hopf algebra.*

As an example, let X be a connected topological group, with $m: X \times X \to X$ the group multiplication map and $\Delta: X \to X \times X$ the diagonal map. Let F be any field. Then $H_*(X;F)$ is a Hopf algebra with multiplication m_* and diagonal map Δ_*. Also, $H^*(X;F)$ is a Hopf algebra, with multiplication Δ^* and diagonal map m^*. Since F is a field, $H^*(X;F)$ is the vector-space dual to $H_*(X;F)$, and in the duality, m_* corresponds to m^* and Δ_* to Δ^*. In a moment we will discuss such duality in the general setting.

At the beginning of this chapter we gave the conditions for the multiplication φ to be associative or commutative. Dualizing, ψ is said to be *associative* if the following diagram is commutative:

$$\begin{array}{ccc} A & \xrightarrow{\psi} & A \otimes A \\ {\scriptstyle\psi}\downarrow & & \downarrow{\scriptstyle\psi\otimes 1} \\ A \otimes A & \xrightarrow{1\otimes\psi} & A \otimes A \otimes A \end{array}$$

while ψ is *commutative* if $T \circ \psi = \psi : A \to A \otimes A$, that is, if the following diagram is commutative:

$$\begin{array}{ccc} & & A \otimes A \\ & {\scriptstyle\psi}\nearrow & \\ A & & \downarrow{\scriptstyle T} \\ & {\scriptstyle\psi}\searrow & \\ & & A \otimes A \end{array}$$

We may interpret Theorem 2 as asserting that the Steenrod algebra $\mathcal{A}$ is a Hopf algebra, since the diagonal map ψ given in Theorem 2 has the required properties. The multiplication of $\mathcal{A}$ is associative but not commutative. However, its co-multiplication is both associative and commutative; we leave the easy verification to the reader. (Since ψ is an algebra homomorphism, it suffices to check on the generators.)

THE DUAL OF THE STEENROD ALGEBRA

Now assume that R is a field, and let A be a (connected) Hopf algebra over R, of finite type (that is, each A_i is finite-dimensional over R). We define the *dual Hopf algebra* to A, A^*, by setting $(A^*) = (A_i)^*$, that is, the dual to A_i as a vector space over R. The multiplication φ in A gives the diagonal map φ^* of A^*, and the diagonal map ψ of A gives the multiplication ψ^* of A^*. It is easy to verify that A^* is a Hopf algebra. Moreover, φ^* is associative if and only if φ is associative; and similarly for φ^* commutative, and for ψ^* associative or commutative. We note that A and A^* are isomorphic as R-modules, but certainly not as algebras in general.

Let $\mathcal{A}^*$ denote the dual to the Steenrod algebra. Then $\mathcal{A}^*$ is a Hopf algebra with associative diagonal map φ^* and with associative and commutative multiplication ψ^*. It will turn out that ψ^* is particularly simple: we will prove that as an algebra $\mathcal{A}^*$ is a polynomial ring.

For convenience we now consider infinite sequences of integers. Let $\mathfrak{R}$ denote the set of all infinite sequences of non-negative integers with only finitely many non-zero entries. Such a sequence will be called *admissible* if it consists of an admissible sequence in our previous sense, followed by all zeroes. Precisely, $I = \{i_1, i_2 \dots, i_r, \dots\}$ is admissible if, for some $r \geq 0$, $i_r > 0$, $i_q \geq 2i_{q+1}$ for $1 \leq q < r$, and $i_s = 0$ for $s > r$. Let $\mathfrak{J}$ denote the subset of $\mathfrak{R}$ consisting of all admissible sequences.

For each integer $k \geq 0$, let I^k be the admissible sequence $\{2^{k-1}, \dots, 2, 1, 0, \dots\}$ (I^0 denotes the zero sequence).

Lemma 1

Let x denote the generator of $H^1(Z_2, 1; Z_2)$. Then $Sq^I(x) = x^{2^k}$ if $I = I^k$, and $Sq^I(x) = 0$ for all other non-zero admissible sequences I. Further, for any non-zero I (admissible or not), $Sq^I(x) = 0$ unless I is obtained from some I^k by interspersion of zeros.

This lemma follows immediately from the observation that $Sq(x^{2^j}) = (Sq\, x)^{2^j} = (x + x^2)^{2^j} = x^{2^j} + x^{2^{j+1}} = Sq^0 x^{2^j} + Sq^{2^j} x^{2^j}$.

Definition

For each $i \geq 0$, let ξ_i be the element of $\mathcal{A}^*_{2^i-1}$ characterized by $\xi_i(\theta)(x)^{2^i} = \theta(x) \in H^{2^i}(Z_2, 1; Z_2)$ for all $\theta \in \mathcal{A}_{2^i-1}$ (ξ_0 is the unit of $\mathcal{A}^*$).

It will sometimes be convenient to write $\langle \xi_i, \theta \rangle$ for $\xi_i(\theta)$.

Proposition 1

Let I be admissible. For $k \geq 1$, $\langle \xi_k, Sq^I \rangle = 1$ if $I = I^k$. Otherwise, $\langle \xi_k, Sq^I \rangle = 0$. Further, for arbitrary I, $\langle \xi_k, Sq^I \rangle = 0$ unless I is obtained from I^k by interspersion of zeros.

This follows immediately from Lemma 1. (We remark that this property could be taken as the definition of ξ_k.)

Now define a set isomorphism $\gamma : \mathfrak{J} \to \mathfrak{R}$ by $\gamma(\{a_1, \dots, a_k, 0, \dots\}) = \{a_1 - 2a_2, a_2 - 2a_3, \dots, a_k, 0, \dots\}$.

For each $R \in \mathfrak{R}$, we define $\xi^R \in \mathcal{A}^*$ by $\xi^R = \prod_{i=1}^{\infty} (\xi_i)^{r_i}$ where $R = \{r_1, r_2, \dots\}$. Note that for $I \in \mathfrak{J}$, the degree of Sq^I is exactly the degree of $\xi^{\gamma(I)}$.

We order the sequences of $\mathfrak{J}$ lexicographically from the right. Thus, for example, $\{8,4,2,0,\dots\} > \{9,4,1,0,\dots\} > \{9,4,0,\dots\} > \{17,3,0,\dots\}$.

Theorem 3

For $I, J \in \mathfrak{J}$, $\langle \xi^{\gamma(J)}, Sq^I \rangle = 0$, $I < J$; if $I = J$, $\langle \xi^{\gamma(J)}, Sq^I \rangle = 1$.

We will prove this by a downward induction. It is trivial for $J = \{0, \dots\}$. Let $J = \{a_1, \dots, a_k, 0 \dots\}$ and $I = \{b_1, \dots, b_k, 0, \dots\}$ where, assuming $J \geq I$, we have $a_k \geq b_k > 0$, $a_k \geq 1$. Put

$$J' = \{a_1 - 2^{k-1}, a_2 - 2^{k-2}, \dots, a_k - 1, 0, \dots\}$$

Then $\xi^{\gamma(J)} = \xi^{\gamma(J')}\xi_k$, since $\gamma(J') = \gamma(J)$ except in the kth place. Calculating, we have

$$\langle \xi^{\gamma(J)}, Sq^I \rangle = \langle \xi^{\gamma(J')}\xi_k, Sq^I \rangle = \langle \psi^*(\xi^{\gamma(J')} \otimes \xi_k), Sq^I \rangle$$
$$= \langle \xi^{\gamma(J')} \otimes \xi_k, \psi(Sq^I) \rangle$$

Using the definition of the diagonal map ψ in $\mathcal{A}$,

$$\langle \xi^{\gamma(J)}, Sq^I \rangle = \langle \xi^{\gamma(J')} \otimes \xi_k, \sum Sq^{I_1} \otimes Sq^{I_2} \rangle$$
$$= \sum \langle \xi^{\gamma(J')}, Sq^{I_1} \rangle \langle \xi_k, Sq^{I_2} \rangle$$

where the summation is over sequences I_1, I_2 (not necessarily admissible) such that $I_1 + I_2 = I$ (in the sense of termwise addition).

Now if $b_k = 0$, I_2 has 0 in the kth place and thereafter; hence, by the above proposition, $\langle \xi_k, Sq^{I_2} \rangle = 0$. If $b_k \neq 0$, we see that the only nonzero term in the above summation occurs for $I_2 = I^k$. Thus $\langle \xi^{\gamma(J)}, Sq^I \rangle = \langle \xi^{\gamma(J')}, Sq^{I'} \rangle$ where $I' = I - I^k$; for this is the only non-zero term in the above summation.

Descent on b_k and k completes the proof.

Corollary 2

As an algebra, $\mathcal{A}^*$ is the polynomial ring over Z_2 generated by the $\{\xi_i\}$ $(i \geq 1)$.

PROOF: The admissible sequences give a vector-space basis for $\mathcal{A}$. As J runs through $\mathfrak{J}$, $\xi^{\gamma(J)}$ runs through all the monomials in the ξ_i. But the above theorem shows that the $\xi^{\gamma(J)}$ (J admissible) form a vector-space basis for $\mathcal{A}^*$. However, a polynomial ring is characterized by the fact that the monomials in the generators form a vector-space basis. This proves the corollary.

ALGEBRAS OVER A HOPF ALGEBRA

Let A and M be a graded R-algebra and a graded R-module, respectively, where R is a commutative ring with unit. We say that M is a *graded A-module* if we are given an R-module homomorphism $\mu: A \otimes M \to M$ such that $\mu(1 \otimes m) = m$ and $\mu \circ (\varphi_A \otimes 1) = \mu \circ (1 \otimes \mu): A \otimes A \otimes M \to M$, where φ_A is the multiplication in A.

Now suppose that A is a Hopf algebra and that M is an A-module (as above) which is also an R-algebra. Then $M \otimes M$ has a natural structure as an A-module under the composition

$$\mu': A \otimes M \otimes M \xrightarrow{\psi \otimes 1 \otimes 1} A \otimes A \otimes M \otimes M \xrightarrow{T}$$
$$A \otimes M \otimes A \otimes M \xrightarrow{\mu \otimes \mu} M \otimes M$$

(where ψ is the diagonal map of A). We say that M is an *algebra over the Hopf algebra A* if the multiplication $\varphi_M\colon M \otimes M \to M$ is a homomorphism of A-modules, that is,

$$\begin{array}{ccc} A \otimes M \otimes M & \xrightarrow{\mu'} & M \otimes M \\ {\scriptstyle 1 \otimes \varphi_M}\downarrow & & \downarrow{\scriptstyle \varphi_M} \\ A \otimes M & \xrightarrow{\mu} & M \end{array}$$

is a commutative diagram.

As an example, take $R = Z_2$, $A = \mathcal{A}$, and $M = H^*(X;Z_2)$ where X is any space. Because of the formal relationship between the Cartan formula and the diagonal map ψ of $\mathcal{A}$, we have, for $\theta \in \mathcal{A}$ and $x,y \in H^*(X;Z_2)$, $\theta(\varphi_M(x \otimes y)) = \theta(xy) = \varphi_M(\psi(\theta)(x \otimes y))$. Thus the cohomology (with Z_2 coefficients) of any space is an algebra over the Hopf algebra $\mathcal{A}$.

As another example, let A be any Hopf algebra, and let B be a graded A-module. Then the tensor algebra $\Gamma(B)$ has a natural structure as an algebra over the Hopf algebra A; we leave the verification to the reader.

THE DIAGONAL MAP OF $\mathcal{A}^*$

In what follows we will write H_* and H^* for $H_*(X;Z_2)$ and $H^*(X;Z_2)$, respectively. The space X will be assumed to be a complex of finite type. Thus we have duality: $(H_*)^* = H^*$, etc. We have a diagonal map $\Delta\colon X \to X \times X$, and the induced maps $\Delta_*\colon H_* \to H_* \otimes H_*$ and $\Delta^* = (\Delta_*)^*\colon H^* \otimes H^* \to H^*$. Recall that H^* is an algebra over the Hopf algebra $\mathcal{A}$, by the diagram (1), where the composition along the top row is μ', which

$$\begin{array}{ccccccc} \mathcal{A} \otimes H^* \otimes H^* & \xrightarrow{\psi} & \mathcal{A} \otimes \mathcal{A} \otimes H^* \otimes H^* & \xrightarrow{T} & \mathcal{A} \otimes H^* \otimes \mathcal{A} \otimes H^* & \xrightarrow{\mu \otimes \mu} & H^* \otimes H^* \\ {\scriptstyle 1 \otimes \Delta^*}\downarrow & & & & & & \downarrow{\scriptstyle \Delta^*} \\ \mathcal{A} \otimes H^* & & & \xrightarrow{\mu} & & & H^* \end{array} \tag{1}$$

makes $H^* \otimes H^*$ an $\mathcal{A}$-module.

Define $\lambda\colon H_* \otimes \mathcal{A} \to H_*$ by $\langle \lambda(x,\theta), y \rangle = \langle x, \mu(\theta,y) \rangle$ for $y \in H^*$, $x \in H_* = (H^*)^*$, and $\theta \in \mathcal{A}$.

Proposition 2

λ is a module operation; i.e., the following diagram commutes.

$$\begin{array}{ccc} H_* \otimes \mathcal{A} \otimes \mathcal{A} & \xrightarrow{\lambda \otimes 1} & H_* \otimes \mathcal{A} \\ {\scriptstyle 1 \otimes \varphi}\downarrow & & \downarrow{\scriptstyle \lambda} \\ H_* \otimes \mathcal{A} & \xrightarrow{\lambda} & H_* \end{array}$$

PROOF: The fact that μ is a module operation, $\mu \circ (1 \otimes \mu) = \mu \circ (\varphi \otimes 1)$, which can be expressed by a diagram closely related to the above, is the essential step in the following calculation:

$$\begin{aligned}
\langle \lambda \circ (1 \otimes \varphi)(x,\theta,\theta'), y \rangle &= \langle \lambda(x,\varphi(\theta,\theta')), y \rangle \\
&= \langle x, \mu(\varphi(\theta,\theta'), y) \rangle \\
&= \langle x, \mu(\theta,\mu(\theta,'y)) \rangle \\
&= \langle \lambda(x,\theta), \mu(\theta',y) \rangle \\
&= \langle \lambda(\lambda(x,\theta),\theta'), y \rangle \\
&= \langle \lambda \circ (\lambda \otimes 1)(x,\theta,\theta'), y \rangle
\end{aligned}$$

Proposition 3

The following diagram commutes:

$$\begin{array}{ccccc}
H_* \otimes H_* \otimes \mathcal{A} \otimes \mathcal{A} & \xrightarrow{T} & H_* \otimes \mathcal{A} \otimes H_* \otimes \mathcal{A} & \xrightarrow{\lambda \otimes \lambda} & H_* \otimes H_* \\
{\scriptstyle \Delta_* \otimes \psi}\uparrow & & & & \uparrow{\scriptstyle \Delta_*} \\
H_* \otimes \mathcal{A} & & \xrightarrow{\qquad\quad \lambda \qquad\quad} & & H_*
\end{array}$$

PROOF: This diagram is related to diagram (1) as the diagram for λ is to that for μ (in Proposition 2), and diagram (1) plays a similar essential role in the proof, which proceeds by the following calculation, where we write $\Delta_* x = \sum x_i' \otimes x_i''$ and $\psi(\theta) = \sum \theta_j' \otimes \theta_j''$:

$$\begin{aligned}
\langle \Delta_* \circ \lambda(x,\theta), y \otimes y' \rangle &= \langle \lambda(x,\theta), \Delta^*(y \otimes y') \rangle \\
&= \langle x, \mu(\theta,\Delta^*(y \otimes y')) \rangle \\
&= \langle x, \Delta^* \circ (\mu \otimes \mu) \circ T \circ \psi(\theta,y,y') \rangle \qquad \text{by diagram (1)} \\
&= \langle \Delta_* x, (\mu \otimes \mu)(\textstyle\sum \theta_j' \otimes y \otimes \theta_j'' \otimes y') \rangle \\
&= \langle \textstyle\sum x_i' \otimes x_j'', \sum \mu(\theta_j',y) \otimes \mu(\theta_j'',y') \rangle \\
&= \textstyle\sum\sum \langle x_i', \mu(\theta_j',y) \rangle \langle x_i'', \mu(\theta_j'',y') \rangle \\
&= \textstyle\sum\sum \langle \lambda(x_i',\theta_j'), y \rangle \langle \lambda(x_i'',\theta_j''), y' \rangle \\
&= \langle (\lambda \otimes \lambda) \textstyle\sum\sum (x_i' \otimes \theta_j' \otimes x_i'' \otimes \theta_j''), y \otimes y' \rangle \\
&= \langle (\lambda \otimes \lambda) \circ T \circ (\Delta_* \otimes \psi)(x,\theta), y \otimes y' \rangle
\end{aligned}$$

Definition

Let λ^* be the dual of λ; $\lambda^* : H^* \to (H_* \otimes \mathcal{A})^* = H^* \hat{\otimes} \mathcal{A}^*$ where we write $\hat{\otimes}$ as a reminder that infinite sums are allowed.

Proposition 4

λ^* is an algebra homomorphism.

The proof consists of dualizing the diagram of Proposition 3.

Proposition 5

The following diagram is commutative.

$$\begin{array}{ccc} H^* \hat{\otimes} \mathcal{A}^* \hat{\otimes} \mathcal{A}^* & \xleftarrow{\lambda^* \otimes 1} & H^* \hat{\otimes} \mathcal{A}^* \\ {\scriptstyle 1 \otimes \varphi^*} \uparrow & & \uparrow {\scriptstyle \lambda^*} \\ H^* \hat{\otimes} \mathcal{A}^* & \xleftarrow{\quad \lambda^* \quad} & H^* \end{array}$$

This is the dual of Proposition 2.

Proposition 6

For $y \in H^*$, the following formulas are equivalent:

1. $\lambda^*(y) = \sum y_i \otimes w_i$ $\quad$ ($y_i \in H^*$, $w_i \in \mathcal{A}^*$; sum possibly infinite)
2. $\mu(\theta,y) = \sum \langle \theta, w_i \rangle y_i$ $\quad$ for all $\theta \in \mathcal{A}$

PROOF: Assume (1). For $x \in H_*$,

$$\begin{aligned} \langle x, \mu(\theta,y)\rangle = \langle \lambda(x,\theta), y\rangle &= \langle x \otimes \theta, \lambda^* y\rangle \\ &= \langle x \otimes \theta, \textstyle\sum y_i \otimes w_i\rangle \\ &= \textstyle\sum \langle x,y_i\rangle\langle \theta,w_i\rangle \\ &= \langle x, \textstyle\sum \langle \theta,w_i\rangle y_i\rangle \end{aligned}$$

which implies (2) by the duality of H_* and H^*. Conversely, assuming (2),

$$\begin{aligned} \langle x \otimes \theta, \lambda^* y\rangle &= \langle x, \mu(\theta,y)\rangle \\ &= \langle x, \textstyle\sum \langle \theta,w_i\rangle y_i\rangle \\ &= \textstyle\sum \langle x,y_i\rangle\langle \theta,w_i\rangle \\ &= \langle x \otimes \theta, \textstyle\sum y_i \otimes w_i\rangle \end{aligned}$$

which implies (1).

Proposition 7

Let x denote the generator of $H^1(Z_2,1:Z_2)$. Then $\lambda^*(x) = \sum_{i \geq 0} x^{2^i} \otimes \xi_i$.

PROOF: By Proposition 6, we must show $\mu(Sq^I,x) = \sum \langle \xi_i, Sq^I\rangle x^{2^i}$, and it is enough to check for admissible I. Now $\langle \xi_i, Sq^I\rangle = 1$ when $I = I^i = \{2^{i-1}, 2^{i-2}, \ldots, 2,1,0,\ldots\}$ and has value 0 otherwise. This means that $\sum_i \langle \xi_i, Sq^I\rangle x^{2^i} = 0$ except for the particular I above, for which the summation becomes just x^{2^i}. But on the other hand $\mu(Sq^I,x) = Sq^I x$, and the theorem follows immediately, by Lemma 1.

Corollary 3

$\lambda^*(x^{2^i}) = \sum_{j \geq 0} x^{2^{i+j}} \otimes (\xi_j)^{2^i}$.

PROOF: From Propositions 4 and 7,

$$\lambda^*(x^{2^i}) = (\lambda^* x)^{2^i} = \left(\textstyle\sum x^{2^j} \otimes \xi_j\right)^{2^i} = \textstyle\sum_j (x^{2^j} \otimes \xi_j)^{2^i} = \textstyle\sum_j x^{2^{i+j}} \otimes (\xi_j)^{2^i}$$

Theorem 4

The diagonal map φ^* of $\mathcal{A}^*$ is given by the formula

$$\varphi^*(\xi_k) = \sum_{i=0}^{k}(\xi_{k-i})^{2^i} \otimes \xi_i$$

PROOF: Using the above formulas, we have

$$(1 \otimes \varphi^*)\lambda^*(x) = (1 \otimes \varphi^*)(\sum_k x^{2^k} \otimes \xi_k) = \sum_k x^{2^k} \otimes \varphi^*(\xi_k)$$

and also

$$(\lambda^* \otimes 1)\lambda^*(x) = (\lambda^* \otimes 1)(\sum_i x^{2^i} \otimes \xi_i) = \sum\sum x^{2^{i+j}} \otimes \xi_j^{2^i} \otimes \xi_i$$

By Proposition 5, these expressions are equal. The theorem follows.

THE MILNOR BASIS FOR $\mathcal{A}$

Recall that $\{\xi^R\}$, $R \in \mathcal{R}$, forms a basis for $\mathcal{A}^*$. The dual basis, whose elements are denoted $\{Sq^R\}$, is called the *Milnor basis* for the Steenrod algebra $\mathcal{A}$. By definition, $\langle \xi^R, Sq^{R'} \rangle = 1$ if $R = R'$ and $\langle \xi^R, Sq^{R'} \rangle = 0$ otherwise.

This is a different basis entirely from the Serre-Cartan basis of admissible sequences; some caution is necessary to distinguish, for example, the element $Sq^{(2,1,0,\ldots)}$ of the Milnor basis from the element Sq^2Sq^1 of the Serre-Cartan basis, especially since one tends in the course of long calculations to abbreviate both of these elements (which do not even have the same dimension) to $Sq^{2,1}$. However, we do have the following compatibility.

Proposition 8

$Sq^{(i,0,\ldots)} = Sq^i$.

PROOF: By Theorem 3, for I,J admissible, $\langle \xi^{\gamma(J)}, Sq^I \rangle$ has value 0 for $I < J$ and is equal to 1 for $I = J$. Observe that for $R = \{i,0,\ldots\}$, $\gamma(R) = R$. Then $\xi^{\gamma(R)} = \xi^R$, and so clearly $\langle \xi^R, Sq^i \rangle = 1$. If, on the other hand, J is another sequence of the same degree as R, then $J > R$, so that $\langle \xi^{\gamma(J)}, Sq^i \rangle = 0$. Thus Sq^i is dual to ξ^R and thus $Sq^i = Sq^R$, $R = \{i,0,\ldots\}$, as was to be proved.

We illustrate with some further remarks showing the usefulness of the dual $\mathcal{A}^*$ in the analysis of the Steenrod algebra $\mathcal{A}$. For $t > 0$, let J_t denote the two-sided ideal of $\mathcal{A}^*$ generated by

$$\{(\xi_1)^{2^t}, (\xi_2)^{2^{t-1}}, \ldots, (\xi_i)^{2^{t-i+1}}, \ldots, (\xi_t)^2; \xi_{t+1}, \xi_{t+2}, \ldots\}$$

Define C_t as the quotient $\mathcal{A}^*/J_t$. As an algebra, C_t is the polynomial ring in $\xi_1, \ldots, \xi_t$ truncated by the relations $(\xi_i)^{2^{t-i+1}} = 0$ $(1 \leq i \leq t)$. C_t is clearly finite-dimensional. Now $\varphi^*(J_t) \subset J_t \otimes \mathcal{A}^*$, since

$$\varphi^*(\xi_k)^{2^{t-k+1}} = \sum_i (\xi_{k-i})^{2^{t-k+i+1}} \otimes (\xi_i)^{2^{t-k+1}}$$

which lies in $J_t \otimes \mathcal{A}^*$. Therefore φ^* defines a map $C_t \rightarrow C_t \otimes C_t$. Thus C_t is actually a quotient Hopf algebra of $\mathcal{A}^*$.

Note also that J_t is maximal among ideals J of $\mathcal{A}^*$ having the following properties:

1. $\xi_1^i \in J$ if and only if $i \geq 2^t$
2. $\varphi^*(J) \subset J \otimes \mathcal{A}^*$, and hence $\mathcal{A}^*/J$ is a quotient Hopf algebra of $\mathcal{A}^*$

Let A_t be the dual of C_t. Then A_t is a finite-dimensional subalgebra of $\mathcal{A}$. Clearly $Sq^i \in A_t$ if and only if $(\xi_1)^i$ is *not* in J_t. Thus every Sq^i is contained in a finite-dimensional subalgebra of $\mathcal{A}$. It follows that every Sq^i is nilpotent. In fact A_t is exactly the subalgebra of $\mathcal{A}$ generated by $\{Sq^i : i < 2^t\}$. Indeed, let B_t be the subalgebra of $\mathcal{A}$ generated by the Sq^i, $i < 2^t$. From the formula for the diagonal map of $\mathcal{A}$, B_t is clearly a sub Hopf algebra of $\mathcal{A}$. Surely $B_t \subset A_t$. But by duality sub Hopf algebras of $\mathcal{A}$ correspond to quotient Hopf algebras of $\mathcal{A}^*$, which in turn are quotients of $\mathcal{A}^*$ by ideals satisfying property (2) above. The result follows by maximality of J_t with respect to properties (1) and (2).

By refinement of the proof of Theorem 1 of Chapter 4, we can easily verify that A_t can also be described as the subalgebra of $\mathcal{A}$ generated by $\{Sq^{2^i} : i \leq t-1\}$. In recent literature the notation has shifted and this subalgebra is often denoted $\mathcal{A}_{t-1}$.

Note finally that $Sq^R \in A_t$ if and only if ξ^R is not in J_t, that is, if and only if we have simultaneously $r_1 < 2^t$, $r_2 < 2^{t-1}, \ldots, r_t < 2$, and $r_k = 0$ for all $k > t$. Thus for every R there exists some t sufficiently large such that Sq^R is in A_t. Thus every element of $\mathcal{A}$ is nilpotent (except of course $Sq^0 = 1$).

DISCUSSION

What we have done in this chapter is to construct an algebraic system with the properties of the Steenrod squaring operations built in. The Adem relations are imposed in the definition of the Steenrod algebra $\mathcal{A}$. The Cartan formula appears in more subtle fashion. Formally, the Cartan formula suggests the diagonal map which makes $\mathcal{A}$ into a Hopf algebra; however, the force of the Cartan formula is that the cohomology ring, under cup product, of a space is then an algebra over the *Hopf* algebra $\mathcal{A}$.

The algebra $\mathcal{A}$ is of course very complicated. The calculation of the Serre-Cartan basis was an important step but left many questions about $\mathcal{A}$ open. Milnor's study of $\mathcal{A}$, based on the dual algebra, yielded a new basis and many new results. However, work in homotopy theory, especially by Adams, has shown that much more deep algebraic information about $\mathcal{A}$ will have geometrical application. In Chapter 18 especially we will illustrate this; let it suffice now to say that $\mathcal{A}$ remains an important area of study.

EXERCISES

1. Using the Serre-Cartan basis and the Adem relations, calculate the subalgebra of $\mathcal{A}$ generated (as an algebra) by Sq^0, Sq^1, and Sq^2.
2. Write the elements in Exercise 1 in terms of the Milnor basis.

REFERENCES

General

1. J. Milnor and J. C. Moore [1].
2. N. E. Steenrod and D. B. A. Epstein [1].

The structure of the Steenrod algebra and the dual algebra

1. J. F. Adams [1].
2. H. Cartan [6].
3. J. Milnor [2].

CHAPTER 7

EXACT COUPLES AND SPECTRAL SEQUENCES

In Chapters 2 through 6 we have considered an explicit family of cohomology operations, namely, the Steenrod squares. As we indicated in Chapter 1, we will also wish to make some explicit computations of $H^*(\pi,n; G)$. For this purpose, and others, we stop to introduce some algebraic machinery—that of the spectral sequence. Our exposition is based on the notion of exact couple.

EXACT COUPLES

An *exact couple* consists of a pair of modules D,E together with homomorphisms i,j,k such that the following triangular diagram is exact at

$$\begin{array}{ccc} D & \xrightarrow{i} & D \\ {\scriptstyle k}\nwarrow & & \swarrow{\scriptstyle j} \\ & E & \end{array}$$

each vertex. If we write $d=jk\colon E \to E$, then d is a differential on E, since $dd=j(kj)k=0$.

Given an exact couple as above, we define the *derived couple* as follows:

$$\begin{array}{ccc} D' & \xrightarrow{i'} & D' \\ {\scriptstyle k'}\nwarrow & & \swarrow{\scriptstyle j'} \\ & E' & \end{array}$$

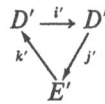

$D' = iD \subset D$; $E' = H(E,d)$; $i' = i \mid D'$; $j' = j \circ i^{-1}$; that is, j' takes $i(x)$ into the d-homology class $[j(x)]$, $x \in D$, and k' takes $[y]$ into $k(y)$, $y \in E$. In order that this make sense, certain verifications are necessary: that $j(x)$ is a cycle and that $j'(z)$, $z \in D'$, depends only on z; that $k(y)$ depends only on $[y]$ and is in the submodule D' of D. We leave these verifications to the reader, along with the proof of the following proposition, which consists of six easy exercises in diagram-chasing.

Proposition 1

The derived couple is exact.

We may therefore repeat the process of derivation, obtaining an infinite sequence of exact couples. The nth derived couple is

$$\begin{array}{ccc} D^n & \xrightarrow{i^n} & D^n \\ & {\scriptstyle k^n}\nwarrow \quad \swarrow{\scriptstyle j^n} & \\ & E^n & \end{array}$$

where D^n is clearly a submodule of the original D and $i^n = i \mid D^n$. The description of E^n, j^n, and k^n is more complicated. E^n is a subquotient of E^{n-1}, namely, $H(E^{n-1},d^{n-1})$ where $d^r = j^r k^r$. Let us consider d^n. First of all, $d^1 = d' = j'k' : E' \to E'$ takes the d-homology class $[y]$ into the class $[j(x)]$ where x satisfies $i(x) = k(y)$. (Such x exists because y is a cycle, $d(y) = 0$; hence $k(y) \in \ker j = \operatorname{im} i$.) Now d^2 is defined on $E^2 = H(E',d')$; if $d'[y] = 0$, $[y] \in E^1 = E'$, then $d^2 = j^2k^2$ takes $[\![y]\!] = [[y]] \in E^2$ into $[j'(x)] \in E^2$ where $x \in D'$ satisfies $i'(x) = k'[y]$ (such x exists because $d'[y] = 0$). Thus $d^2[\![y]\!] = [j'(i')^{-1}k'[y]]$ where it is defined. If we pursue this analysis, we find that d^3 is defined when d^1 and d^2 are zero and is equal to $j'(i')^{-1}(i')^{-1}k'$; and in general d^n is defined when $d^1,\ldots,d^{n-1}$ are all zero and is equal to $j'(i')^{-(n-1)}k'$. (In the last expression, the superscript denotes an iteration, and should not be confused with the indexing superscripts, as in i^n.)

When we are given an exact couple $(D,E;\, i,j,k)$ and form the sequence of derived couples $(D^n,E^n;\, i^n,j^n,k^n)$, each E^n is a differential group with differential $d^n = j^nk^n$ and $E^{n+1} = H(E^n,d^n)$. The sequence (E^n,d^n) is called the *spectral sequence* of the exact couple; in general, the term "spectral sequence" denotes a sequence of differential groups with the property $E^{n+1} = H(E^n,d^n)$.

THE BOCKSTEIN EXACT COUPLE

An important example is the Bockstein exact couple, where i^1 is induced by multiplication by 2 in Z; j^1 is induced by the reduction homomorphism

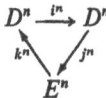

$$\begin{array}{ccc} D^1 = H^*(\ ;Z) & \xrightarrow{i^1} & H^*(\ ;Z) \\ & {}_{k^1}\nwarrow \quad \swarrow_{j^1} & \\ & E^1 = H^*(\ ;Z_2) & \end{array}$$

$\rho: Z \rightarrow Z_2$; and k^1 is the Bockstein homomorphism β (see Chapter 3). The differential $d = j^1k^1$ of this exact couple is just the Bockstein homomorphism δ_2 (see Chapter 3). We will take the above exact couple as the first in a sequence and consider the derived couple E^2, etc. For later convenience we will use subscripts (rather than superscripts) to index the successive Bockstein differentials d_r; thus $d_1 = d$ as above; $d_2 = j^2k^2: E^2 \rightarrow E^2$; etc.

The operation d_r acts essentially as follows: take a cocycle in Z_2 coefficients; represent it by an integral cocycle; take its coboundary; divide by 2^r (this is possible because d_r is defined only on the kernel of d_{r-1}); and reduce the coefficients mod 2. Notice that every d_r raises dimension by 1 in $H^*(\ ;Z_2)$.

The Bockstein differentials act on cohomology with Z_2 coefficients, but they can be made to give information about integral cohomology. To see this, let X be a space and recall that

$$H^p(X;G) = \text{Hom}\,(H_p(X),G) \oplus \text{Ext}\,(H_{p-1}(X),G)$$

by the universal coefficient theorem. Therefore a direct summand Z in $H_p(X;Z)$ gives rise to a summand Z in $H^p(X;Z)$ and to a summand Z_2 in $H^p(X;Z_2)$. A summand Z_{2^n} in $H_p(X;Z)$ gives rise to a Z_{2^n} in $H^{p+1}(X;Z)$ and to summands Z_2 in $H^p(X;Z_2)$ and $H^{p+1}(X;Z_2)$. The following proposition is now self-evident.

Proposition 2

Elements of $H^*(X;Z_2)$ which come from *free* integral classes lie in $\ker d_r$ for every r and not in $\operatorname{im} d_r$ for any r. If z generates a cyclic direct summand of order 2^r in $H^{n+1}(X;Z)$, then there exist cyclic direct summands of order 2 in $H^n(X;Z_2)$ and $H^{n+1}(X;Z_2)$, generated, say, by z' and z'', respectively; $d_i(z')$ and $d_i(z'')$ are zero for $i < r$, and $d_r(z') = z''$. (Implicitly, z' and z'' are also not in $\operatorname{im} d_i$ for $i < r$.)

We say that the image by ρ of the free subgroup of $H^*(X;Z)$ "persists to E^∞" and that z' and z'' "persist to E^r but not to E^{r+1}."

The following application will be important in our computations of the homotopy groups of spheres.

Corollary 1

Suppose that for some X, $H^i(X;Z_2) = 0$, $i < n$, and $H^n(X;Z_2) = Z_2$, generated, say, by z. Then we can infer $H^n(X;Z)$, except for torsion in odd primes, as follows: $H^n(X;Z)=Z$ if $d_r z=0$ for all r; and $H^n(X;Z)=Z_{2^n}$ if $d_i z=0$, $i<n$, and $d_n z\neq 0$.

THE SPECTRAL SEQUENCE OF A FILTERED COMPLEX

We turn now to a detailed study of another example: the exact couple of a filtered complex, and the associated spectral sequence.

Let (K,d) be a chain complex. Let $\{K^p\}$ be a filtration of K, that is, a family of subcomplexes of K such that $K^p\subset K^{p+1}$ for every p. Later we will assume that $K^p=0$ for $p<0$ and that $\bigcup_p K^p=K$, but for the present these hypotheses are not necessary.

From the short exact sequence

$$0\longrightarrow K^{p-1}\longrightarrow K^p\longrightarrow K^p/K^{p-1}\longrightarrow 0$$

we obtain a long exact sequence in homology,

$$\cdots\xrightarrow{\partial} H_{p+q}(K^{p-1})\xrightarrow{i} H_{p+q}(K^p)\xrightarrow{j} H_{p+q}(K^p/K^{p-1})\longrightarrow H_{p+q-1}(K^{p-1})\longrightarrow\cdots \qquad (1)$$

where the coefficient group, understood to be fixed, is suppressed from the notation. We thus obtain a bigraded exact couple by taking

$$\begin{aligned}
&D_{p,q}=H_{p+q}(K^p) \qquad E_{p,q}=H_{p+q}(K^p/K^{p-1})\\
i_{p,q}&:\ D_{p,q}\longrightarrow D_{p+1,q-1}:\ H_{p+q}(K^p)\longrightarrow H_{p+q}(K^{p+1})\\
j_{p,q}&:\ D_{p,q}\longrightarrow E_{p,q}:\ H_{p+q}(K^p)\longrightarrow H_{p+q}(K^p/K^{p-1})\\
k_{p,q}&:\ E_{p,q}\longrightarrow D_{p-1,q}:\ H_{p+q}(K^p/K^{p-1})\longrightarrow H_{p+q-1}(K^{p-1})
\end{aligned}$$

so that the portion of the long exact sequence (1) displayed above becomes, in the notation of the exact couple,

$$\cdots\xrightarrow{k_{p,q+1}} D_{p-1,q+1}\xrightarrow{i_{p-1,q+1}} D_{p,q}\xrightarrow{j_{p,q}} E_{p,q}\xrightarrow{k_{p,q}} D_{p-1,q}\xrightarrow{i_{p-1,q}}\cdots$$

Finally, the differential of the exact couple is

$$d=jk:\ E_{p,q}\longrightarrow E_{p-1,q}:\ H_{p+q}(K^p/K^{p-1})\longrightarrow H_{p+q-1}(K^{p-1}/K^{p-2})$$

In this exact couple, the groups D and E are bigraded; the bidegrees of the maps i,j,k,d are implicit in the representations given above. In Figure 1 we display a portion of the diagram in which all these groups and maps appear. The reader should trace the long exact sequence (1) through this diagram and should also see how the differential d appears there.

$$
\begin{array}{ccccccccc}
 & & & & & & \cdots \to H_{p+q-1}(K^{p-r}) & \xrightarrow{j} & H_{p+q-1}(K^{p-r}/K^{p-r-1}) \\
 & & & & & & \downarrow{\scriptstyle i} & & \\
 & & & & & & \cdots \to H_{p+q-1}(K^{p-r+1}) & \xrightarrow{j} & H_{p+q-1}(K^{p-r+1}/K^{p-r}) \\
 & & \downarrow & & & & \wr\downarrow{\scriptstyle i^{(r-3)}} & & \cdots\cdots\cdots\cdots \\
\to H_{p+q+1}(K^{p}/K^{p-1}) & \xrightarrow{k_{p,q+1}} & H_{p+q}(K^{p-1}) & \xrightarrow{j} & H_{p+q}(K^{p-1}/K^{p-2}) & \xrightarrow{k} & H_{p+q-1}(K^{p-2}) & \xrightarrow{j} & H_{p+q-1}(K^{p-2}/K^{p-3}) \\
 & & \downarrow{\scriptstyle i_{p-1,q+1}} & & & & \downarrow{\scriptstyle i} & & \\
\to H_{p+q+1}(K^{p+1}/K^{p}) & \xrightarrow{k} & H_{p+q}(K^{p}) & \xrightarrow{j_{p,q}} & H_{p+q}(K^{p}/K^{p-1}) & \xrightarrow{k_{p,q}} & H_{p+q-1}(K^{p-1}) & \xrightarrow{j} & H_{p+q-1}(K^{p-1}/K^{p-2}) \\
(= E_{p+1,q}) & & (= D_{p,q}) & & (= E_{p,q}) & & (= D_{p-1,q}) & & (= E_{p-1,q}) \\
 & & \downarrow{\scriptstyle i_{p,q}} & & & & \downarrow{\scriptstyle i_{p-1,q}} & & \\
\to H_{p+q+1}(K^{p+2}/K^{p+1}) & \xrightarrow{k} & H_{p+q}(K^{p+1}) & \xrightarrow{j} & H_{p+q}(K^{p+1}/K^{p}) & \xrightarrow{k} & H_{p+q-1}(K^{p}) & \xrightarrow{j} & H_{p+q-1}(K^{p}/K^{p-1}) \\
\cdots\cdots\cdots\cdots & & \wr\downarrow{\scriptstyle i^{(r-3)}} & & & & \downarrow{\scriptstyle i} & & \\
\to H_{p+q-1}(K^{p+r-1}/K^{p+r-2}) & \xrightarrow{k} & H_{p+q}(K^{p+r-2}) \to \cdots & & & & & &
\end{array}
$$

Figure 1

Now attach a superscript 1 to D,E,i,j,k,d in the above, and consider the derived couple. We have

$$D^2_{p,q} = \text{im } i\colon D^1_{p-1,q+1} \to D^1_{p,q}$$

$$E^2_{p,q} = H(E^1_{p,q}, d^1) = \frac{\ker d^1_{p,q} : E^1_{p,q} \to E^1_{p-1,q}}{\text{im } d^1_{p+1,q} : E^1_{p+1,q} \to E^1_{p,q}}.$$

As before, the maps i^2, etc., are given subscripts to agree with those of the domain; thus

$$i^2_{p,q}\colon D^2_{p,q} \to D^2_{p+1,q-1}$$

and so forth. The reader should verify that in the nth derived couple the bidegrees of the maps are as follows:

MAP	BIDEGREE
i^n	$(1, -1)$
j^n	$(-n+1, n-1)$
k^n	$(-1,0)$
$d^n = j^n k^n$	$(-n, n-1)$

Observe how d^r appears in Figure 1. In the first place, $d^1_{p,q}$ goes from the group $E_{p,q}$ (in the center of Figure 1) two steps to the right, by k and then j. If an element is in the kernel of d^1, $d^1(x) = 0$ $(x \in E^1_{p,q})$, then $k(x)$ is in the image of the appropriate i, by exactness of (1); and $d^2(x)$ is essentially the class obtained by taking x "right one, up one, right one." We have remarked before that d^2 is essentially $j^1(i^1)^{-1}k^1$. In general d^{r+1} is essentially $j(i)^{-r}k$, which on the diagram means "right one, up r, right one"—this making sense if d^r vanishes.

We have the following formula for $E^r_{p,q}$.

Proposition 3

$E^r_{p,q}$ is isomorphic to the quotient group

$$\frac{\text{im}: H_{p+q}(K^p/K^{p-r}) \to H_{p+q}(K^p/K^{p-1})}{\text{im } \partial : H_{p+q+1}(K^{p+r-1}/K^p) \to H_{p+q}(K^p/K^{p-1})}$$

For convenience we shall abbreviate this quotient to "A/B." The map ∂ which is used in B is the boundary homomorphism of the exact homology sequence of the triple $(K^{p+r-1}, K^p, K^{p-1})$, while the map of which A is the image is one of the maps in the exact sequence of the triple (K^p, K^{p-1}, K^{p-r}). Note that, by exactness of this last sequence, $A = \ker \partial\colon H_{p+q}(K^p/K^{p-1}) \to H_{p+q-1}(K^{p-1}/K^{p-r})$, where ∂ is the boundary homomorphism of the triple in question.

PROOF: If $r=1$, the proposition reduces to a triviality. If $r=2$, we have $A=\ker \partial: H_{p+q}(K^p/K^{p-1}) \to H_{p+q-1}(K^{p-1}/K^{p-2})$, but this map is exactly the map $d^1_{p,q}$; moreover, B is just $\operatorname{im} d^1_{p+1,q}$; thus the proposition is obvious in this case also.

Now suppose $r>2$. We seek an isomorphism $\bar{f}: A/B \to E^r_{p,q}$. We first note that there is a natural homomorphism $f: A \to E^{r-1}_{p,q}$. To see this, suppose $x \in A$ ($=\ker \partial$ as above). This means that x is an element of $H_{p+q}(K^p/K^{p-1}) = E^1_{p,q}$ such that $\partial(x) = 0 \in H_{p+q-1}(K^{p-1}/K^{p-r})$. Thus $k_{p,q}(x)$ is in the image of $i^{(r-1)}$: $H_{p+q-1}(K^{p-r}) \to H_{p+q-1}(K^{p-1})$ (use Figure 1, where these groups and maps are displayed). Then $i^{-(r-2)}k(x)$ is in im $i=\ker j$, so that $d^{r-1}(x)=0$. Thus we have shown not only that x projects to $E^{r-1}_{p,q}$ but that it projects into $\ker d^{r-1}_{p,q}$. We therefore have a homomorphism $f: A \to Z^{r-1}_{p,q}$ where

$$Z^{r-1}_{p,q} = \ker d^{r-1}_{p,q}: E^{r-1}_{p,q} \to E^{r-1}_{p-r+1,q+r-2}$$

In order to show that f induces a map from A/B to $E^r_{p,q}$, it is now sufficient to show that $f(B) \subset B^{r-1}_{p,q}$ where

$$B^{r-1}_{p,q} = \operatorname{im} d^{r-1}_{p+r-1,q-r+2}: E^{r-1}_{p+r-1,q-r+2} \to E^{r-1}_{p,q}$$

But this is clear; for, if $x \in B$, then $x = \partial(y)$ for some

$$y \in H_{p+q+1}(K^{p+r-1}/K^p)$$

Let y' be the image of y in $H_{p+q+1}(K^{p+r-1}/K^{p+r-2})$—the latter group appears in the lower left corner of Figure 1; then $f(x) = d^{r-1}(y')$.

Thus we have a natural homomorphism

$$\bar{f}: A/B \to E^r_{p,q} = Z^{r-1}_{p,q}/B^{r-1}_{p,q}$$

To see that $\bar{f}$ is onto, let x be an arbitrary element of $Z^{r-1}_{p,q}$. Choose $x' \in E^1_{p,q}$ such that x' projects onto x. Then $k(x') = i^{(r-2)}(y)$ for some $y \in H_{p+q-1}(K^{p-r+1})$ and we can assume $j(y)=0$. But $j(y)=0$ implies $y \in \operatorname{im} i$, or $k(x') \in \operatorname{im}(i^{(r-1)})$, so that $x' \in A = \ker \partial: H_{p+q}(K^p/K^{p-1}) \to H_{p+q-1}(K^{p-1}/K^{p-r})$ and $f(x') = x$. This proves that $\bar{f}$ is onto.

To see that $\bar{f}$ is 1-1, suppose $f(x) \in B^{r-1}_{p,q}$; we must show $x \in B$. Now $x = d^{r-1}(y)$ for some $y \in H_{p+q+1}(K^{p+r-1}/K^{p+r-2})$. Since $k(y)$ is in the image of $i^{(r-2)}$, if we consider the triple $(K^{p+r-1}, K^{p+r-2}, K^p)$, we have $\partial(y)=0$ in $H_{p+q}(K^{p+r-2}/K^p)$, and so there exists $z \in H_{p+q+1}(K^{p+r-1}/K^p)$, which goes to y in the exact sequence of that triple. Denoting by w the image of z in $H_{p+q}(K^p)$, we have $i^{(r-2)}(w) = k(y)$; hence $x = d^{r-1}(y) = j(w) = \partial(z)$, so that $x \in B$.

This completes the proof of Proposition 3.

Definition

$F_{p,q} = \text{im}: H_{p+q}(K^p) \to H_{p+q}(K)$.

Definition

$E^{\infty}_{p,q} = F_{p,q}/F_{p-1,q+1}$.

The second definition makes sense because $F_{p-1,q+1} = \text{im}: H_{p+q}(K^{p-1}) \to H_{p+q}(K)$ and this latter map can be factored through $H_{p+q}(K^p)$.

The $F_{p,q}$ form a filtration of $H_*(K)$ and the $E^{\infty}_{p,q}$ form the associated graded group. There is no *a priori* connection between E^r and E^{∞}, but the notation is suggestive, and the following formula, analogous to Proposition 3, reinforces the suggestion.

Proposition 4

$E^{\infty}_{p,q}$ is isomorphic to the quotient group

$$\frac{\text{im}: H_{p+q}(K^p) \to H_{p+q}(K^p/K^{p-1})}{\text{im}\ \partial: H_{p+q+1}(K/K^p) \to H_{p+q}(K^p/K^{p-1})}$$

where the maps in question come from the homology exact sequences of the pair (K^p,K^{p-1}) and the triple (K,K^p,K^{p-1}), respectively.

We shall abbreviate this quotient as "C/D".

The proof follows immediately from the diagram below and a standard argument (see Exercise 1).

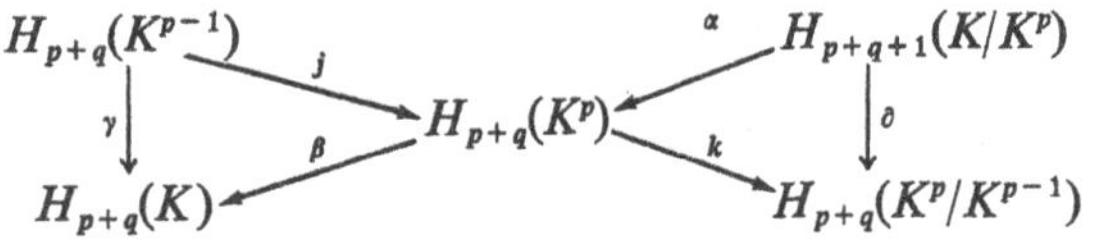

Our intention is to use the spectral sequence to get information about $H_*(K)$ from knowledge of the subcomplexes $\{K^p\}$. In order to press this program, we now consider the following *convergence conditions* on the subcomplexes $\{K^p\}$:

1. $K^p = 0$ for all $p < 0$
2. $E^1_{p,q} = H_{p+q}(K^p/K^{p-1}) = 0$ for all $q < 0$
3. $K = \bigcup_p K^p$

It follows from (1) that $E^1_{p,q} = 0$ for all $p < 0$, and thus it follows from (1) and (2) that $E^r_{p,q} = 0$ for all r whenever $p < 0$ or $q < 0$. The next two results show how much more we can deduce from conditions (1) to (3).

Proposition 5

If (1) and (2) hold, then $E^r_{p,q} = E^{r+1}_{p,q}$ for any $r > \max(p, q+1)$.

The proof is easy: by definition, $E^{r+1}_{p,q}$ is the homology of $E^r_{p,q}$ with respect to d^r as follows:

$$E^r_{p+r,q-r+1} \xrightarrow{d^r} E^r_{p,q} \xrightarrow{d^r} E^r_{p-r,q+r-1}$$

but when $r > \max(p, q+1)$, the two groups at the ends of this sequence are both zero, as a consequence of (1) and (2).

Theorem 1

If convergence conditions (1) to (3) hold and $r > \max(p, q+1)$, then $E^r_{p,q} \approx E^\infty_{p,q}$. Thus the convergence conditions imply that, for each (p,q), the groups $E^r_{p,q}$ become stable with sufficiently large r and the stable group is exactly $E^\infty_{p,q}$.

PROOF: Recall from Propositions 3 and 4 that $E^r_{p,q} \approx A/B$ and $E^\infty_{p,q} \approx C/D$. (For the definitions of A, B, C, D, refer to Propositions 3 and 4.)

We first show that $B = D$, provided $r > q + 1$. In the diagram below,

$$\begin{array}{ccccc} H_{p+q+1}(K^{p+r-1}/K^p) & \xrightarrow{\varphi} & H_{p+q+1}(K^{p+r}/K^p) & \rightarrow & H_{p+q+1}(K^{p+r}/K^{p+r-1}) \\ & \searrow^{\partial} & \swarrow_{\partial'} & & \\ & H_{p+q}(K^p/K^{p-1}) & & & \end{array}$$

the row is exact, and the last group in the row is $E^1_{p+q,q-r+1} = 0$ by condition (2), so that φ is onto and thus $\operatorname{im} \partial = \operatorname{im} \partial'$. Under hypothesis (3), it follows that $B = D$.

If $r > p$, then (1) implies that $H_{p+q}(K^p/K^{p-r}) = H_{p+q}(K^p)$. Thus we have $A = C$. Therefore when $r > p$ and $r > q + 1$, we have $A/B = C/D$, and the theorem is proved.

Recall that $E^\infty_{p,q} = F_{p,q}/F_{p-1,q+1}$ where $F_{p,q} = \operatorname{im}: H_{p+q}(K^p) \rightarrow H_{p+q}(K)$. Assuming the convergence conditions (1) to (3), we have $E^\infty_{p,q} = 0$ for $p < 0$ or $q < 0$. Therefore the following series is finite:

$$H_n(K) = F_{n,0} \supset F_{n-1,1} \supset \cdots \supset F_{1,n-1} \supset F_{0,n}$$

and the successive quotients are the groups $E^\infty_{i,n-i}$. This composition series gives us considerable information on $H_n(K)$; we say "the spectral sequence converges to $H_*(K)$." Note for fixed n, only finitely many computations are needed to obtain a composition series for $H_n(K)$.

EXAMPLE: THE HOMOLOGY OF A CELL COMPLEX

As an application, let K be the singular chain complex of a cell complex (also denoted by K) and K^p be the singular chain complex of the p-skeleton; H_* will denote singular homology (with some fixed coefficient group).

The convergence conditions (1) to (3) are clearly satisfied; moreover we have a strong form of condition (2), since $E^1_{p,q} = H_{p+q}(K^p/K^{p-1}) = 0$ for all $q \neq 0$. It follows from this last observation that $E^r_{p,q} = 0$ for $q \neq 0$, and $E_{p,q}$ likewise. The composition series collapses, and we have

$$H_n(K) = F_{n,0} = E_{n,0}$$

Also, d^r must be identically zero for $r > 1$, since it changes the second grading; thus $H_n(K) = E_{n,0} = E^1_{n,0}$. From the definitions we easily see that $H_n(K)$ is nothing but the homology at $H_p(K^p/K^{p-1})$ in the following long sequence (not exact),

$$\begin{array}{ccccc} \cdots \to H_{p+1}(K^{p+1}/K^p) & \xrightarrow{d^1} & H_p(K^p/K^{p-1}) & \xrightarrow{d^1} & H_{p-1}(K^{p-1}/K^{p-2}) \to \cdots \\ = E^1_{p+1,0} & & = E^1_{p,0} & & = E^1_{p-1,0} \end{array}$$

where each d^1 is the boundary homomorphism of a certain triple of the form (K^p, K^{p-1}, K^{p-2}). Thus we have proved the familiar result that the singular homology of the cell complex K may be computed as the homology of a complex $\bar{C}(K)$, in which $\bar{C}_p(K) = H_p(K^p/K^{p-1})$.

DOUBLE COMPLEXES

An important source of filtered chain complexes is the double complex. A *double complex* is a doubly indexed family $\{C_{p,q}\}$ of abelian groups, with two differentials,

$$d' : C_{p,q} \to C_{p-1,q}, \qquad d'' : C_{p,q} \to C_{p,q-1}$$

such that $d'd' = 0$, $d''d'' = 0$, and $d'd'' + d''d' = 0$. We also assume that $C_{p,q} = 0$ for $p < 0$ or $q < 0$.

An example of a double complex is given by setting $C_{p,q} = A_p \otimes B_q$, where (A, d_A) and (B, d_B) are given chain complexes; we put $d' = d_A \otimes 1_B$, $d'' = (-1)^p 1_A \otimes d_B : A_p \otimes B_q \to A_p \otimes B_{q-1}$.

A double complex $(C_{p,q}, d', d'')$ gives rise to a chain complex (C_n, d) if we set $C_n = \sum_{p+q=n} C_{p,q}$ and $d = d' + d''$. There are two obvious filtrations of the complex (C_n, d):

1. $'C^p_n = \sum_{\substack{j+q=n \\ j \leq p}} C_{j,q}$

2. $''C^q_n = \sum_{\substack{p+k=n \\ k \leq q}} C_{p,k}$

Each of these filtrations gives rise to a spectral sequence. For instance, in (1), one has $'E^1_{p,q} = H_{p+q}('C^p/'C^{p-1})$. Here $('C^p/'C^{p-1})_n = C_{p,n-p}$ or, equivalently, $('C^p/'C^{p-1})_{p+q} = C_{p,q}$. One verifies that the above homology group $'E^1$ is computed by means of d'' and that the differential d^1 is induced by d'. Thus, in a rough but natural notation, we may write $'E^2_{p,q} = H'_pH''_q(C)$. In (2) the situation is similar but the roles of the two indices are interchanged. We write $''E^2_{p,q} = H''_qH'_p(C)$.

Each filtration obviously satisfies the convergence conditions, and therefore each spectral sequence converges to the same thing, namely, $H_*(C)$.

We illustrate these ideas with a theorem in homological algebra. We recall the definition of Tor (A,B) for abelian groups A,B. In the first place, we may define 'Tor (A,B) by taking a free abelian resolution

$$0 \to R' \to F' \to A \to 0$$

of A; 'Tor (A,B) is characterized by the exact sequence

$$0 \to \text{'Tor}\,(A,B) \to R' \otimes B \to F' \otimes B \to A \otimes B \to 0$$

In the second place, we may define "Tor (A,B) by a resolution of B

$$0 \to R'' \to F'' \to B \to 0$$

and the requirement that the following be exact:

$$0 \to \text{"Tor}\,(A,B) \to A \otimes R'' \to A \otimes F'' \to A \otimes B \to 0$$

From either definition, it follows that Tor $(A,B) = 0$ if either A or B is free.

Proposition 6

'Tor = "Tor.

PROOF: Let X be the chain complex with $X_1 = R'$, $X_0 = F'$, $X_i = 0$ otherwise. Let Y be the complex $Y_1 = R''$, $Y_0 = F''$, $Y_i = 0$ otherwise. Let C be the double complex $X \otimes Y$. Then

$$[C_{p,q}] = \begin{bmatrix} F' \otimes R'' & R' \otimes R'' \\ F' \otimes F'' & R' \otimes F'' \end{bmatrix}$$

We first compute $H_*(C)$ from the spectral sequence of the filtration (1). First,

$$H''_q(C) = \begin{bmatrix} \text{"Tor}\,(F',B) & \text{"Tor}\,(R',B) \\ F' \otimes B & R' \otimes B \end{bmatrix} = \begin{bmatrix} 0 & 0 \\ F' \otimes B & R' \otimes B \end{bmatrix}$$

and finally,

$$H'_p(H''_q(C)) = \begin{bmatrix} 0 & 0 \\ A \otimes B & \text{'Tor}\,(A,B) \end{bmatrix}$$

This is E^2 and hence also E^∞. Thus $H_0(C) = A \otimes B$ and $H_1(C) = {}'\mathrm{Tor}(A,B)$. On the other hand, computing from the spectral sequence of (2),

$$H'_p(C) = \begin{bmatrix} A \otimes R'' & {}'\mathrm{Tor}\,(A,R'') = 0 \\ A \otimes F'' & {}'\mathrm{Tor}\,(A,F'') = 0 \end{bmatrix}$$

$$H''_q(H'_p(C)) = \begin{bmatrix} {}''\mathrm{Tor}\,(A,B) & 0 \\ A \otimes B & 0 \end{bmatrix}$$

By this route we obtain $H_0(C) = A \otimes B$, but $H_1(C) = {}''\mathrm{Tor}\,(A,B)$. Thus ${}'\mathrm{Tor}\,(A,B) = H_1(C) = {}''\mathrm{Tor}\,(A,B)$.

We have thus also proved that Tor (A,B) is independent of the choice of free abelian resolution.

DISCUSSION

We have chosen to emphasize the exact couple as the starting point for the theory of spectral sequences. This point of view, due to Massey, seems conceptually simpler than earlier methods, which were based directly on differential filtered groups, or than the more abstract methods of Cartan and Eilenberg.

On the other hand, while we have developed the spectral sequence of a filtered complex as an example of a bigraded exact couple, we now note (with Cartan-Eilenberg) that the results obtained do not really depend on the existence of the complex K. To construct the exact couple, we need only a family of groups fitting into the diagram of Figure 1, that is, playing the roles of $H_*(K^p)$ and $H_*(K^{p+1}/K^p)$. If we have as well groups playing the roles of $H_*(K)$, $H_*(K/K^p)$, and $H_*(K^{p+r}/K^p)$—that is, fitting into the relevant exact sequences—and satisfying suitable convergence conditions, an analogue of Theorem 1 will apply. The reader will find examples in the Appendix below and in Chapter 14.

APPENDIX: THE HOMOTOPY EXACT COUPLE

As another example we describe the *homotopy exact couple*. The results include a derivation of a certain exact sequence of J. H. C. Whitehead, who discovered it by other methods. For this example we assume the relative Hurewicz theorem.

Let K be a CW-complex; consider the exact homotopy sequence of the pair (K^p,K^{p-1}). By setting

$$D^1_{p,q} = \pi_{p+q}(K^p) \qquad E^1_{p,q} = \pi_{p+q}(K^p,K^{p-1})$$

we obtain the homotopy exact couple, with a diagram analogous to Figure 1 of this chapter. (We remark first that E^1 is not a homotopy invariant of K and second that we leave to the reader the definitions of $D^1_{p,q}$ and $E^1_{p,q}$ for small p and q.) The convergence conditions will clearly be satisfied if we have the relative Hurewicz theorem for K (so that $E^1_{p,q} = 0$ for $q < 0$); hence we assume K simply connected. Then this exact couple gives a spectral sequence which converges to $\pi_*(K)$. Consider the derived couple (which is a homotopy invariant, but we leave the proof to the reader); a typical portion of this couple appears as follows:

$$\cdots \longrightarrow E^2_{p+1,0} \xrightarrow{k^2} D^2_{p,0} \xrightarrow{i^2} D^2_{p+1,-1} \xrightarrow{j^2} E^2_{p,0} \longrightarrow \cdots \tag{1}$$

It can be shown that $E^2_{p,0} \approx H_p(K)$. In fact it is clear that $E^1_{p,0} \approx C_p(K)$; and one verifies that d^1 is the same as the homology boundary. Thus the above sequence (1) contains the homology of K. Moreover, $D^2_{p+1,-1} = \operatorname{im} j\colon \pi_p(K^p) \longrightarrow \pi_p(K^{p+1}) = \pi_p(K)$, but j is onto, so that the sequence (1) also contains the homotopy of K. If we denote by $\Gamma_p(K)$ the group $D^2_{p,0} = \operatorname{im} j\colon \pi_p(K^{p-1}) \longrightarrow \pi_p(K^p)$, then (1) becomes the exact sequence of J. H. C. Whitehead:

$$\cdots \longrightarrow H_{p+1}(K) \longrightarrow \Gamma_p(K) \longrightarrow \pi_p(K) \longrightarrow H_p(K) \longrightarrow \cdots$$

As a corollary, suppose K is $(n-1)$-connected. Then the inclusion $K^{n-1} \subset K^n$ is null-homotopic, by an easy obstruction-theory argument; hence $\Gamma_p(K) = 0$ for $p \leq n$, and hence, if K is $(n-1)$-connected, then the Hurewicz homomorphism $\pi_{n+1}(K) \longrightarrow H_{n+1}(K)$ is *onto*.

EXERCISES

1. Prove the following "butterfly lemma": Given a commutative diagram

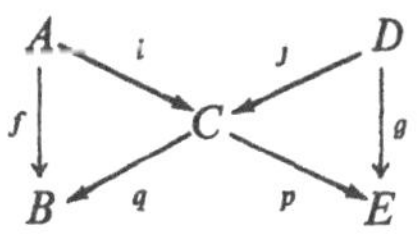

of abelian groups and homomorphisms, in which the diagonals pi and qj are exact at C, there is an isomorphism

$$\frac{\operatorname{im} q}{\operatorname{im} f} \approx \frac{\operatorname{im} p}{\operatorname{im} g}$$

(Proposition 4 is a special case.)

2. Prove that the homotopy exact couple is not a homotopy invariant but that the first derived couple is (see Appendix).

REFERENCES

Expositions of spectral sequences

1. H. Cartan [6].
2. — and S. Eilenberg [1].
3. R. Godement [1].
4. S.-T. Hu [1].
5. W. S. Massey [1,2].
6. J.-P. Serre [1].

Appendix

1. P. J. Hilton [1].
2. J. H. C. Whitehead [2].

CHAPTER 8

FIBRE SPACES

In this chapter we recall elementary properties of fibre spaces and state properties of Serre's spectral sequence for the homology of a fibre space.

FIBRE SPACES

Let E and B be topological spaces, and let p be a map of E onto B. The map p is a *fibre map* in the sense of Serre if it satisfies the covering homotopy condition for finite complexes; that is, given any finite complex K, a map $g\colon K \to E$, and a map $F\colon K \times I \to B$ such that $F(x,0) = p(g(x))$ for all $x \in K$ (thus F is a homotopy of pg), then there exists a map $G\colon K \times I \to E$ such that $G(x,0) = g(x)$ and such that $p(G(x,t)) = F(x,t)$. (Thus G is a homotopy of g which covers the homotopy F.)

In the same context, p is a fibre map in the sense of Hurewicz if it satisfies the covering homotopy condition for all topological spaces; i.e., if in the detailed definition above, K is an arbitrary topological space. Clearly every Hurewicz fibre map is also a Serre fibre map; the converse is false.

We say that (E,p,B) is a *fibre space* if p is a fibre map.

We will always assume that B, the "base space," is arc-connected. (This makes redundant the assumption that p be onto B.) And when the sense is not specified, "fibre map" will be understood to mean: in the sense of Serre.

Choose a base point $*$ in B. Then $p^{-1}(*)$ is a subspace of E, called the *fibre* F of the fibre space. For any $b \in B$, the subspace $p^{-1}(b)$ of E is called the *fibre over* b. In a Hurewicz fibre space, all the fibres have the same homotopy type; in a Serre fibre space, any two fibres which are both finite complexes must have the same homotopy type.

A fibre map p induces, in homotopy, an isomorphism

$$p_*: \pi_*(E,F) \approx \pi_*(B)$$

and, using this isomorphism, the exact homotopy sequence of the pair (E,F) becomes the *exact homotopy sequence of the fibre space:*

$$\cdots \longrightarrow \pi_n(F) \longrightarrow \pi_n(E) \longrightarrow \pi_n(B) \longrightarrow \pi_{n-1}(F) \longrightarrow \cdots$$

We assume that the reader is familiar with the general theory of fibre spaces, including the notions set forth above, as well as certain others, in particular the action of $\pi_1(B)$ on the fibre and the coefficient bundle $\mathcal{H}_q(F)$ over B. (Although results will often be stated in terms of (co)homology with local coefficients, in practice we will assume that the local coefficients are simple.) However, we will further discuss the general theory of fibre spaces later, especially in Chapters 11 and 14.

We assemble several important examples.

Example 1

(R,p,S^1) where R is the real line and S^1 may be considered as the reals modulo 1, the interval $I = [0,1]$ with the endpoints identified; p is the natural projection. The fibre is Z, that is, an infinite discrete group which we identify with the integers. From the exact homotopy sequence of this fibre space we deduce the homotopy groups of S^1:

$$\pi_1(S^1) \approx Z \qquad \pi_n(S^1) = 0 \text{ for } n > 1$$

Example 2

(S^3,η,S^2) where η is the Hopf map. The fibre is S^1. From the previous example, $\pi_n(S^1) = 0$ for $n > 1$, and therefore the exact homotopy sequence of the Hopf fibration takes the form

$$\cdots \longrightarrow 0 \longrightarrow \pi_3(S^3) \longrightarrow \pi_3(S^2) \longrightarrow 0 \longrightarrow \pi_2(S^3) \longrightarrow \pi_2(S^2) \longrightarrow Z \longrightarrow 0$$

which confirms that $\pi_2(S^2) \approx Z$ and shows that, for all $n > 2$, $\pi_n(S^3) \approx \pi_n(S^2)$. Thus in particular $\pi_3(S^2) = Z$ generated by the class of η.

Example 3

(E,p,B) where B is an arc-connected space with base point $*$, E is the space of paths in B beginning at $*$, and p projects a path onto its terminal point. The path space E is given the compact-open topology. It is easy to show that E is contractible; the usual contracting map pulls each path back along itself to the base point. The fibre is ΩB, the space of loops in B at $*$; since E is contractible, the exact homotopy sequence shows that, for any space B,

$$\pi_n(\Omega B) \approx \pi_{n+1}(B)$$

Example 4

(C,p,B) where C is a regular covering space of B and p is the covering map, which is a local homeomorphism. The fibre G is a discrete group, the group of covering translations. Since G is discrete, we have, from the exact sequence, $\pi_n(C) \approx \pi_n(B)$ for $n > 1$, whereas the exact sequence terminates as follows:

$$\pi_1(G) \to \pi_1(C) \xrightarrow{p_*} \pi_1(B) \to G \to \pi_0(C)$$

and the groups on the ends are trivial, so that $G \approx \pi_1(B)/p_*\pi_1(C)$. (Since C is a regular covering, $\pi_1(C)$ is a normal subgroup.) If C is simply connected (the universal covering), then $G \approx \pi_1(B)$.

Example 5

$(S^{2n-1}, p, CP(n-1))$, the base space being complex projective space; the fibre is S^1; it is understood that $n > 1$. From the exact sequence, we have $\pi_2(CP(n-1)) \approx \pi_1(S^1) \approx Z$, and $\pi_i(CP(n-1)) \approx \pi_i(S^{2n-1})$ for $i \geq 2$. Thus the first non-trivial homotopy groups of $CP(n-1)$ are infinite cyclic groups in dimensions 2 and $2n-1$. It follows that $CP(\infty)$ is a $K(Z,2)$ space.

Example 6

$(\mathbf{O}(n+1), p, S^n)$ where the total space $\mathbf{O}(n+1)$ is the orthogonal group in $n+1$ variables and p is evaluation at a fixed point of S^n. The fibre is $\mathbf{O}(n)$. From the exact sequence

$$\pi_{i+1}(S^n) \to \pi_i(\mathbf{O}(n)) \to \pi_i(\mathbf{O}(n+1)) \to \pi_i(S^n)$$

we have, if $n > i+1$, $\pi_i(\mathbf{O}(n)) \approx \pi_i(\mathbf{O}(n+1))$, so that $\pi_i(\mathbf{O}(n))$ is independent of n for n sufficiently large $(n > i+1)$. This limiting group is called the stable group $\pi_i(\mathbf{O})$. These groups are known; by the Bott periodicity theorem, $\pi_i(\mathbf{O}) \approx \pi_{i+8}(\mathbf{O})$, $i \geq 0$. The values are

i mod 8	0	1	2	3	4	5	6	7
$\pi_i(\mathbf{O})$	Z_2	Z_2	0	Z	0	0	0	Z

THE HOMOLOGY SPECTRAL SEQUENCE OF A FIBRE SPACE

Fibre spaces may be thought of as generalized product spaces. In homotopy, the analogy with product spaces is expressed through the exact homotopy sequence of the fibre space. In homology the situation is more complicated. The main results express the analogy between the total space E and the product space $B \times F$ by means of a spectral sequence which, under certain hypotheses, starts from the product of the homology of B and of F and converges to the homology of E.

These results are obtained by means of cubical homology theory, or, to be precise, based normalized cubical singular homology theory. We will not derive them. The main theorem is as follows.

Theorem 1 (Serre)

Let (E,p,B) be a fibre space with fibre F, and suppose that B and F are arcwise connected.

1. The chain complex $C(E)$ admits a certain filtration, giving rise to a spectral sequence.
2. In the resulting spectral sequence $\{E^r,d^r\}$, the bidegree of d^r is $(-r, r-1)$.
3. $E^1_{p,q} = C_p(B) \otimes H_q(F)$.
4. $E^2_{p,q} = H_p(B;\mathcal{H}_q(F))$—local coefficients. If B is simply connected, then $E^2_{p,q} = H_p(B;H_q(F))$.
5. The spectral sequence converges to $H_*(E)$.
6. $i_*: H_n(F) \to H_n(E)$ may be computed as follows:
   $H_n(F) = E^2_{0,n} \to E^r_{0,n} = E^\infty_{0,n} = F_{0,n} \subset H_n(E)$
   r being taken sufficiently large so that $E^r_{0,n} = E^\infty_{0,n}$. Note that every d^r vanishes on $E^r_{0,n}$; hence the map $E^2_{0,n} \to E^r_{0,n}$ is defined.
7. $p_*: H_n(E) \to H_n(B)$ may be computed as follows:
   $H_n(E) = F_{n,0} \to E^\infty_{n,0} = E^r_{n,0} \subset E^2_{n,0} = H_n(B)$
   where again r is taken sufficiently large. The existence of an inclusion $E^r_{n,0} \subset E^2_{n,0}$ follows from the obvious fact that no element of $E^r_{n,0}$ can be a boundary for any r.

EXAMPLE: $H_*(\Omega S^n)$

To illustrate the application of Serre's spectral sequence, we compute $H_*(\Omega S^n)$, the integral homology groups of the loop space of the n-sphere. (We assume $n \geq 2$, so that $B = S^n$ is simply connected.) Recall, from

Example 3 above, that ΩS^n is the fibre of a contractible fibre space over S^n. Hence, in the spectral sequence of this fibre space, $E^\infty_{p,q} = 0$ except for $E^\infty_{0,0} \approx Z$. We know $H_i(S^n) \approx Z$ for $i = 0$ or n, 0 otherwise. It follows from item (4) of the theorem that $E^2_{p,q} = 0$ unless $p = 0$ or n. Thus the only differential in the spectral sequence which can possibly be non-zero is d^n: $E^n_{n,q} \to E^n_{0,q+n-1}$. Consider Figure 1, which displays E^n. We have $E^n_{n,0} \approx Z$.

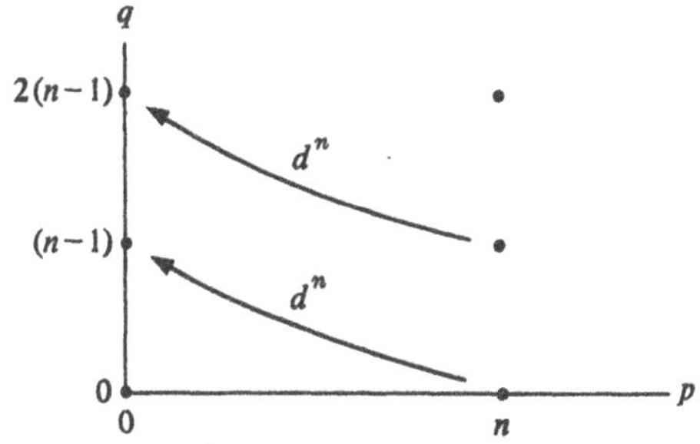

Figure 1

Since $E^{n+1}_{n,0} = E^\infty_{n,0} = 0$, the generator of $E^n_{n,0}$ cannot be a cycle under d^n; moreover, since $E^{n+1}_{0,n-1} = E^\infty_{0,n-1} = 0$, this generator must map $E^n_{n,0}$ isomorphically onto $E^n_{0,n-1}$. This tells us that $H_{n-1}(\Omega S^n) \approx E^2_{0,n-1} = E^n_{0,n-1} \xleftarrow[\approx]{d^n} E^n_{n,0} \approx Z$. Of course we also see that $E^2_{0,q}$ must be zero for $0 < q < n - 1$, and hence $H_q(\Omega S^n)$ likewise. In informal language, the groups $E^2_{0,q}$ for $0 < q < n - 1$ must be zero groups because there is nothing that can "kill" them, and in order to "get rid of" the Z at $(n,0)$ we must have a Z at $(0, n - 1)$.

We now apply similar reasoning to the next $(n - 1)$ rows. From item (4), $E^2_{n,n-1} = H_n(S^n; H_{n-1}(\Omega S^n)) = H_n(S^n; Z) = Z$. To get rid of this Z, we must have an isomorphism d^n: $E^n_{n,n-1} \xrightarrow[\approx]{} E^n_{0,2(n-1)}$, and the inference is that $H_q(\Omega S^n) = 0$ for $(n - 1) < q < 2(n - 1)$, with $H_{2(n-1)} \approx Z$.

Once the pattern is established, it is obvious that an inductive proof gives $H_q(\Omega S^n) \approx Z$ for q an integral multiple of $(n - 1)$, and zero otherwise. This completes the calculation.

SERRE'S EXACT SEQUENCE

Although in general a fibre space does not yield a long exact sequence in homology as it does in homotopy, the Serre spectral sequence gives the following result in that direction.

Theorem 2

Let $(E,p,B;F)$ be a fibre space, with B simply connected. Suppose that $H_i(B)=0$ for $0<i<p$, and that $H_j(F)=0$ for $0<j<q$. Then there is an exact sequence (of finite length),

$$H_{p+q-1}(F)\xrightarrow{i_*}H_{p+q-1}(E)\xrightarrow{p_*}H_{p+q-1}(B)\xrightarrow{\tau}H_{p+q-2}(F)$$
$$\rightarrow\cdots\rightarrow H_1(E)\rightarrow 0$$

PROOF: It follows from the hypotheses that $E^2_{i,j}=0$ when either $0<i<p$ or $0<j<q$. Thus the normal series for $H_n(E)$, which contains the $E^\infty_{i,j}$ terms for which $i+j=n$, collapses to the exact sequence

$$0\rightarrow E^\infty_{0,n}\rightarrow H_n(E)\rightarrow E^\infty_{n,0}\rightarrow 0$$

Now in the general case we have the exact sequence

$$0\rightarrow E^\infty_{n,0}\rightarrow E^n_{n,0}\xrightarrow{d^n}E^n_{0,n-1}\rightarrow E^\infty_{0,n-1}\rightarrow 0$$

When $n<p+q$, we have $E^n_{n,0}\approx H_n(B)$ and $E^n_{0,n-1}\approx H_{n-1}(F)$. Making these substitutions in the last above sequence and splicing together the two types of sequences displayed above (for all $n<p+q$) yields the sequence of the theorem.

We refer to the sequence of the theorem as the *exact homology sequence* of the fibre space, or *Serre's exact sequence* (in homology).

The map $\tau\colon H_n(B)\rightarrow H_{n-1}(F)$, which corresponds to $d^n_{n,0}$, is called the *transgression*. Thus far it is defined only when $n<p+q$ where p,q are as in the hypotheses of the above theorem; later we will study the transgression in greater generality. That the other maps in the exact sequence are actually i_* and p_* is stated in terms (6) and (7) of the main theorem.

Our theory casts new light on the Hurewicz theorem.

Theorem 3 (Hurewicz)

Let X be a connected space with $\pi_i(X)=0$ for all $i<n$ $(n\geq 2)$. Then $H_i(X)=0$ for all $i<n$ and $\pi_n(X)\approx H_n(X)$.

We take as known the corresponding theorem in which $n=1$ and the last statement is replaced by $\pi_1(X)/C\approx H_1(X)$ where C is the commutator subgroup of $\pi_1(X)$. We can then prove the present theorem by induction on n, using Serre's exact sequence. Consider the contractible fibre space $(E,p,X;\,\Omega X)$. By the induction hypothesis, $H_i(X)=0$ for all $i<n-1$, and $H_{n-1}(X)\approx\pi_{n-1}(X)=0$. It remains to show that $H_n(X)\approx\pi_n(X)$. Now $\pi_j(\Omega X)\approx\pi_{j+1}(X)$; so we can apply the induction hypothesis to the space ΩX and we obtain $H_j(\Omega X)=0$ for all $j<n-1$ and $H_{n-1}(\Omega X)\approx\pi_{n-1}(\Omega X)\approx\pi_n(X)$. We apply Serre's exact sequence, with $p=n$ and $q=n-1$;

$p+q-1=2n-2\geq n$, so that $H_n(X)\approx H_{n-1}(\Omega X)$. But $H_{n-1}(\Omega X)\approx \pi_n(X)$. This completes the proof.

(We omit to show that the isomorphism of the present theorem is in fact the usual Hurewicz homomorphism.)

In somewhat similar fashion one can give a proof of the relative Hurewicz theorem. We omit it here, however, because we will state and prove this theorem in greater generality in Chapter 10.

It is well known that any map f: $X\to Y$ is homotopically equivalent to an inclusion map f': $X\to Y'$ where Y' has the homotopy type of Y. (The standard procedure is to take for Y' the mapping cylinder of f.) This lends weight to the following theorem of J. H. C. Whitehead.

Theorem 4

Suppose f: $X\subset Y$, where X is arc-connected, Y simply connected, and $\pi_2(Y,X)$ abelian. Then, for $n\geq 1$, the statement that f_* (the induced map in homology) is isomorphic for dimensions $j<n$ and epimorphic in dimension n is exactly equivalent to the corresponding statement for $f_\#$ (the induced map in homotopy).

The proof is easy by our present methods. Write the homotopy–homology ladder of the pair (Y,X):

$$\begin{array}{ccccccccc}
\pi_n(X) & \xrightarrow{f_\#} & \pi_n(Y) & \to & \pi_n(Y,X) & \to & \pi_{n-1}(X) & \to & \cdots \\
\downarrow & & \downarrow & & \downarrow & & \downarrow & & \\
H_n(X) & \xrightarrow[f_*]{} & H_n(Y) & \to & H_n(Y,X) & \to & H_{n-1}(X) & \to & \cdots
\end{array}$$

Now consider the following statements:

1. $f_\#$ is isomorphic, $j<n$, and epimorphic, $j=n$
2. The pair (Y,X) is n-connected
3. $H_j(Y,X)=0$ for $j\leq n$
4. f_* is isomorphic, $j<n$, and epimorphic, $j=n$

Statements (1) and (2) are equivalent, by the exactness of the homotopy sequence; (3) and (4) are equivalent by the exactness of the homology sequence. Finally, (2) and (3) are equivalent by the relative Hurewicz theorem. Thus (1) and (4) are equivalent, which establishes the theorem.

Corollary 1

If X and Y are simply connected and f: $X\to Y$, then f_* is isomorphic for all n if and only if $f_\#$ is isomorphic for all n.

This follows from the theorem upon observation that if $\pi_1(X)$ is abelian, then $\pi_2(Y,X)$ is abelian (as required in the theorem). Of course Y is replaced by an equivalent space such that f becomes an inclusion; this does not affect the desired conclusion.

THE TRANSGRESSION

We return to the study of the transgression. The transgression has appeared, in the case where the fibre F is $(q-1)$-connected and the base space B is $(p-1)$-connected, as a map $\tau\colon H_n(B) \to H_{n-1}(F)$, defined when $n < p+q$ and given by $d^n_{n,0}$.

We now define the transgression more generally as $d^n_{n,0}$, without supposing anything about connectivity of F and B. Then the transgression τ in general has a subgroup of $H_n(B)$ as its domain and takes values in a quotient group of $H_{n-1}(F)$.

There is an entirely different description of the transgression which will be important in what follows. In the fibre space $(E,p,B;F)$ let p_0 denote the map of pairs induced by $p, p_0\colon (E,F) \to (B,*)$. We define (for any n) a transformation $\bar{\tau}$ from $\operatorname{im}(p_0)_*$, which is a subgroup of $H_n(B,*)$, to a quotient group of $H_{n-1}(F)$ as follows: if $x \in \operatorname{im}(p_0)_* \subset H_n(B,*)$, choose $y \in H_n(E,F)$ such that $(p_0)_*(y) = x$ and take ∂y as a representative of $\bar{\tau}(x)$. The value of $\bar{\tau}(x)$ lies in a quotient group because of the indeterminacy arising from the choice of y.

We say that $x \in H_n(B)$ is *transgressive* if $\tau(x)$ is defined. Identifying $H_n(B)$ with $E^2_{n,0}$, this is equivalent to the condition that $d^i(x) = 0$ for all $i < n$. We assert that this is in turn equivalent to the condition that x lie in $\operatorname{im}(p_0)_*$ and that, moreover, if $x = (p_0)_*(y)$, then $\tau(x)$ is the homology class of ∂y. Thus $\tau = \bar{\tau}$. We omit the proof; the reader is referred to Serre's paper [1].

THE COHOMOLOGY SPECTRAL SEQUENCE OF A FIBRE SPACE

We now outline the situation for cohomology, which is entirely analogous to the situation for homology as described above but is enriched by the ring structure and the squaring operations.

As in Theorem 1, we have a spectral sequence (E_r, d_r) of which $E_2^{p,q}$ is given by $H^p(B;\mathcal{H}^q(F))$ or by $H^p(B; H^q(F))$ if B is simply connected. We have $d_r^{p,q}\colon E_r^{p,q} \to E_r^{p+r,q-r+1}$, so that the bidegree of d_r is $(r, -r+1)$, just the opposite of that of the homology differential d^r. The spectral sequence converges to $H^*(E)$. The reader can translate most of Theorem 1 and the related discussion into the cohomological case.

As for the ring structure, if R is a ring, then we have the following results (we omit the proof).

Theorem 5

There is a spectral sequence with $E_2^{p,q} = H^p(B;\ \mathcal{H}^q(F;R))$, converging to $H^*(E;R)$; moreover:

1. For each r, E_r is a bigraded ring, i.e., the ring multiplication maps $E_r^{p,q} \otimes E_r^{p',q'}$ into $E_r^{p+p',q+q'}$
2. In E_r, d_r is an anti-derivation (with respect to total degree), i.e., the following is true whenever it makes sense:
   $d_r(ab) = d_r(a)\cdot b + (-1)^k a \cdot d_r(b) \qquad k = \text{total degree of } a$
3. The product in the ring E_{r+1} is induced by the product in E_r, and the product in the ring E_∞ is induced by the cup product in $H^*(E;R)$.

If R is a field (and in fact we will usually work with $R = Z_p$ for p prime), then the Künneth theorem implies that

$$E_2 = H^*(B;R) \otimes H^*(F;R) \qquad \text{(tensor product of rings)}$$

Serre's exact sequence in cohomology is exactly analogous to the sequence in homology; if B and F are $(p-1)$- and $(q-1)$-connected, respectively, the sequence terminates as follows:

$$\cdots \to H^{p+q-2}(F) \xrightarrow{\tau} H^{p+q-1}(B) \xrightarrow{p^*} H^{p+q-1}(E) \xrightarrow{i^*} H^{p+q-1}(F)$$

The (cohomology) transgression is given by

$$\tau = d_n^{0,n-1} : E_n^{0,n-1} \to E_n^{n,0}$$

We say that $x \in H^{n-1}(F)$ is *transgressive* if $\tau(x)$ is defined or, equivalently, if $d_i(x) = 0$ for all $i < n$; as before, this turns out to be equivalent to the condition that δx lies in $\operatorname{im} p^* \subset H^n(E,F)$; and, moreover, if $\delta x = p^*(y)$, then $\tau(x)$ contains y.

Proposition 1

If x is transgressive, then so also is $Sq^i(x)$; moreover, if $y \in \tau(x)$, then $Sq^i(y) \in \tau(Sq^i(x))$.

The proof is an immediate consequence of the preceding remarks; in fact, if $y \in \tau(x)$, then $p^*y = \delta x$, so that $Sq^i(p^*y) = Sq^i(\delta x)$. By naturality, $p^*(Sq^i(y)) = \delta(Sq^i(x))$; but this is equivalent to $Sq^i(y) = \tau(Sq^i(x))$.

DISCUSSION

The spectral sequence of a fibre space has been a calculational tool of the utmost importance in the development of homotopy theory. The main results (Theorems 1, 2, and 5) of this chapter were proved by Serre; at the time they represented a major breakthrough in technique. While we have not proved these theorems here, we urge the reader to consult Serre's paper [1] both for the proofs of these theorems and for many other examples and applications.

We call the reader's attention to Figure 1 and suggest that he use such a graph of E^2 (and higher E^r) to facilitate his own computations.

Theorem 3 (the Hurewicz theorem) and Theorem 4 were known before Serre's work; however, use of his results shortens the proofs.

The periodicity theorem of Bott, mentioned in Example 6, has generated much recent interest in several areas of topology and geometry; we do not pursue any of these directions here.

EXERCISE

1. Compute the *ring* $H^*(\Omega S^n; Z)$. Your result should depend on the parity of n.

REFERENCES

Fibre spaces
1. E. Fadell [1].
2. P. J. Hilton [1].
3. S.-T. Hu [1].
4. J.-P. Serre [1].
5. N. E. Steenrod [1].

The spectral sequence of a fibration
1. S.-T. Hu [1].
2. J.-P. Serre [1].

The periodicity theorem
1. R. Bott [1].
2. D. Husemoller [1].

CHAPTER 9

COHOMOLOGY OF $K(\pi,n)$

Since $K(Z,1)$ may be taken to be the circle S^1, its cohomology is completely known. Also, in Proposition 2 of Chapter 2 we showed that the cohomology ring $H^*(Z_2,1;Z_2)$ is just the polynomial ring $Z_2[\alpha]$ where α is the one-dimensional generator. These two examples provide a starting point in the computations of $H^m(\pi,n;G)$ in general, which is equivalent to the classification of cohomology operations of type $(\pi,n;G,m)$ (Chapter 1, Theorem 2). We will here compute $H^*(\pi,n;G)$ in a number of cases.

Our method will be to use the Serre spectral sequence of fibre spaces in which, of the base space, total space, and fibre, two have known cohomology, thus permitting the calculation of the cohomology of the third.

TWO TYPES OF FIBRE SPACES

There are two types of fibre spaces which will be important in this attack. One is the type of Example 3 of Chapter 8: starting from the base space B, we construct $(E,p,B;\Omega B)$ where the total space is the contractible space of paths in B initiating at the base point of B and the fibre is the space of loops in B. Since E is contractible, it follows from the exact homotopy sequence that if $B=K(\pi,n)$, then $F=\Omega B=K(\pi, n-1)$.

The other type originates from a given short exact sequence of abelian groups,

$$0 \to A \to B \to C \to 0$$

Such an exact sequence gives rise to a fibre space for each n. To see this,

$$\begin{array}{ccc} K(A,n) & \xrightarrow{i} & K(B,n) \\ & & \downarrow p \\ & & K(C,n) \end{array}$$

we use the following fact.

Proposition 1

Every map $f: X \to Y$ is homotopically equivalent to a map $f': X' \to Y$ where (X', f', Y) is a fibre space in the sense of Hurewicz. Moreover, X may be taken to be a deformation retract of X'.

PROOF: Take X' to be the subspace $\{(x,\lambda): f(x) = \lambda(0)\}$ of $X \times Y^I$ and put $f'(x, \lambda) = \lambda(1)$. A homotopy equivalence between X and X' is given by the map $\varphi: X \to X': x \to (x, f(x)^*)$, where $f(x)^*$ denotes the constant path at $f(x)$, and by the map $\psi: X' \to X: (x,\lambda) \to x$. Then $\varphi\psi$ is homotopic to the identity map of X' under a homotopy which retracts all paths back to their initial points. Since $\psi\varphi$ is itself the identity map of X, this shows that X is a deformation retract of X'. Clearly $f'\varphi = f$, so that f and f' are homotopically equivalent.

It is not hard to see that X' is a fibre space in the sense of Hurewicz. Let g_0 be a map of a space A into X' and let $\{h_t\}$ be a homotopy of the map $h_0 = f'g_0$. For each $a \in A$, $g_0(a) = (x(a), \lambda(a))$ gives a path $\lambda(a)$ starting at $x(a)$ and ending at $h_0(a)$. It is required to give a homotopy $\{g_t\}$ of the map $g_0: A \to X'$ such that $g_t(a) = (x_t(a), \lambda_t(a))$ projects to $h_t(a)$, which means that the path $\lambda_t(a)$ must end at the point $h_t(a)$. This is easy; we let $x_t(a) = x(a)$ for all t and obtain $\lambda_t(a)$ by attaching to $\lambda(a)$ the path from $h_0(a)$ to $h_t(a)$ given by the homotopy $\{h_t\}$, with a suitable reparametrization. We leave the explicit details to the reader.

Now suppose we are given an exact sequence of groups, as above. We may realize the homomorphism $B \to C$ by a map $K(B,n) \to K(C,n)$, using Corollary 1 of Chapter 1. We convert this map into a fibre map, and it is obvious from the exact homotopy sequence of the fibre space that the fibre has the homotopy groups of $K(A,n)$. What is not obvious is that the fibre has the homotopy type of a CW complex. We omit to prove this fact.

To illustrate the use of the first type of fibration, we calculate $H^*(Z,2; Z)$.

Proposition 2

$H^*(Z,2; Z)$ is the polynomial ring $Z[\iota_2]$ where ι_2 is of degree 2, that is, $(\iota_2)^n$ generates $H^{2n}(Z,2; Z)$.

We use the fibre space

$$\begin{array}{ccc} S^1 = K(Z,1) & \to & E \\ & & \downarrow \\ & & K(Z,2) \end{array}$$

where E is contractible, and consider the cohomology spectral sequence with integral coefficients. Here $E_2^{0,q}=0$ for $q>1$, so that nonzero groups can appear only in rows 0 and 1 of E_2 (and consequently E_r). Thus only d_2 can be non-zero; all higher d_r are automatically zero for dimensional reasons. Moreover $E_\infty^{p,q}=0$ if $(p,q)\neq(0,0)$, since E is contractible. It follows immediately that $H^p(Z,2;\ Z)\approx E_2^{p,0}$ is Z in even dimensions and zero in odd dimensions; see Figure 1. It remains to obtain the ring structure;

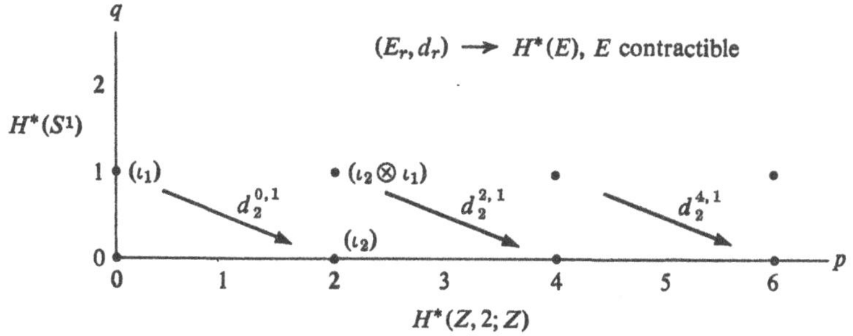

Figure 1 Computing $H^*(Z,2;\ Z)$

for this, we use the fact that d_2 is an anti-derivation and the observation that $d_2^{p,1}$ is an isomorphism for each p and hence carries generators to generators. Thus, denoting the generators of $E_2^{0,1}$ and $E_2^{2,0}$ by ι_1 and ι_2, respectively, we have

$$d_2(\iota_2\otimes\iota_1)=d_2(\iota_2)\otimes\iota_1\pm\iota_2\otimes d_2(\iota_1)=\pm\,\iota_2\otimes\iota_2$$

which shows that $\iota_2\otimes\iota_2$ is a generator of $E_2^{4,0}$, and hence $(\iota_2)^2$ generates $H^4(Z,2;Z)$; similarly,

$$d_2(\iota_2^2\otimes\iota_1)=\iota_2^2\otimes d_2(\iota_1)=\iota_2^2\otimes\iota_2$$

so that $(\iota_2)^3$ generates H^6; and so forth. This proves the proposition. Observe that the cohomology rings of the base space and the fibre are a polynomial algebra and an exterior algebra, respectively.

To illustrate the use of the second type of fibre space, we calculate $H^*(Z_2,1;Z_2)$ directly from $H^*(Z,1)$ without recourse to the construction given in Chapter 2. From the exact sequence

$$0\rightarrow Z\rightarrow Z\rightarrow Z_2\rightarrow 0$$

we construct a fibre space in which both the total space and the fibre have the homotopy type of S^1, while the base space is $K(Z_2,1)$. Although in this case $\pi_1(B)\neq 0$, it can be shown that $\mathcal{H}^*(F)$ is simple. We can then calculate $H^*(Z_2,1;\ Z_2)$, using the spectral sequence of this fibre space (coefficients in Z_2)—see Figure 2. Only two rows can contain non-zero groups, since $F = S^1$. Since $E = S^1$, $E_\infty^{p,q} = 0$ except when $(p+q) = 0$ or 1; moreover, since d_2 is the only non-zero differential in the spectral sequence, it is readily seen that $E_\infty^{0,0} = E_\infty^{1,0} = Z_2$ while $E_\infty^{p,q} = 0$ otherwise. One then deduces $E_2^{p,q}$ $(q = 0,1)$ inductively with increasing p and obtains the desired groups. As for the ring structure, the only point which is not obvious is whether $\alpha^2 \neq 0$ where α generates $H^1(Z_2,1; Z_2)$. Since $\alpha^2 = Sq^1\alpha$ and Sq^1 is the Bockstein homomorphism, it is enough to show that the Bockstein is non-zero on α. We leave this to the reader.

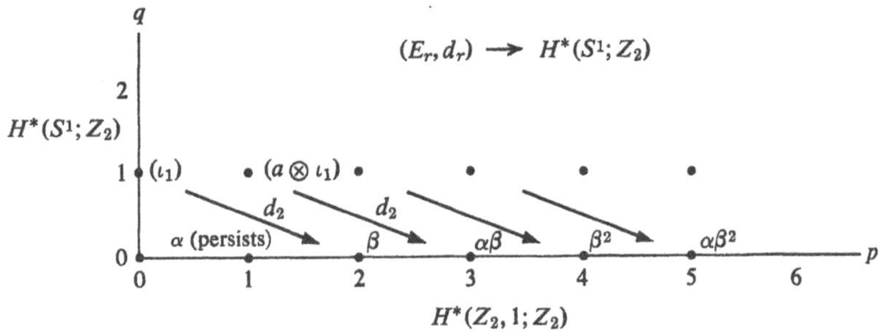

Figure 2 Computing $H^*(Z_2,1; Z_2)$

These examples should convince the reader that skillful use of the spectral-sequence diagram can (in some cases) reduce the calculation to routine. Henceforward we will depend on the reader's ability to work with these diagrams. One generally draws a single diagram and considers it to represent all the E_r superimposed upon one another; only in very complicated cases is it necessary to draw separate diagrams for each E_r.

CALCULATION OF $H^*(Z_2,2; Z_2)$

We now consider a more complicated example: we calculate $H^*(Z_2,2;Z_2)$, at least in low dimensions. For this purpose we use the fibre space over $B = K(Z_2,2)$ with contractible total space and with fibre $F = K(Z_2,1)$. The spectral sequence (with Z_2 coefficients) converges to the cohomology of the total space; hence $E_\infty^{p,q} = 0$ except at (0,0). The cohomology of the fibre we know to be the polynomial ring on a one-dimensional generator α. We proceed as follows with the calculation of the cohomology of the base space—see Figure 3.

By the Hurewicz theorem $H^1(B;Z_2)=0$; hence $E_r^{1,q}=0$ for all q and all r. Then α is transgressive and $\tau(\alpha)=d_2\alpha=\iota_2$, which is the generator of a Z_2 at $E_2^{2,0}$, indicating that $H^2(B;Z_2)=Z_2$. Now $H^2(F;Z_2)$ is generated by α^2, and since d_2 is a derivation, $d_2(\alpha^2)=0$; more generally $d_2(\alpha^{2k})=0$ for all k. On the other hand, $d_2(\alpha^{2k+1})=d_2(\alpha)\otimes\alpha^{2k}=\iota_2\otimes\alpha^{2k}$. Since E_2 is the tensor product of $H^*(B;Z_2)$ and $H^*(F;Z_2)$, $E_2^{2,q}=Z_2$ generated by $\iota_2\otimes\alpha^q$. The groups in this column with q even are killed by $d_2^{0,q+1}$. The others are not cycles; $E_2^{2,2k+1}=Z_2$ generated by $\iota_2\otimes\alpha^{2k+1}$, and $d_2(\iota_2\otimes\alpha^{2k+1})=\iota_2\otimes(\iota_2\otimes\alpha^{2k})=(\iota_2)^2\otimes\alpha^{2k}$. In particular this shows that $E_2^{4,0}$ contains a Z_2 generated by $(\iota_2)^2$. Note that, since ι_2 is of dimension 2, $(\iota_2)^2=Sq^2(\iota_2)$.

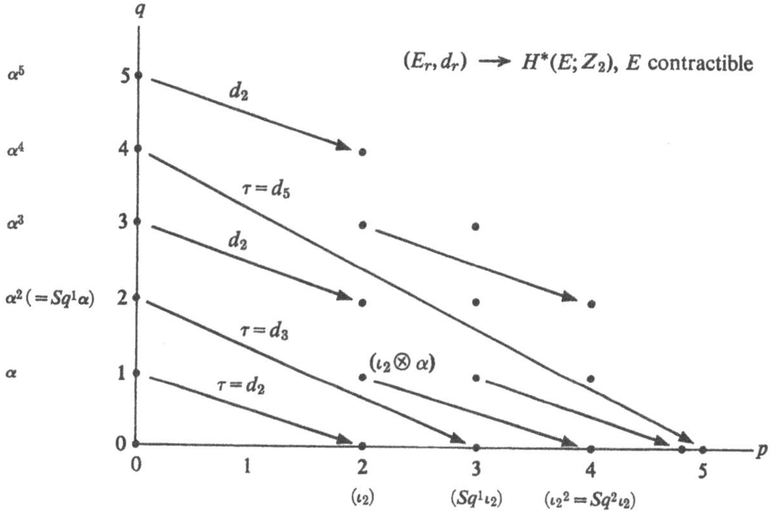

Figure 3 Computing $H^*(Z_2,2;Z_2)$

Thus the column $p=2$ contains groups that alternately kill by or are killed by d_2. We have not yet completed the study of the column $p=0$, where $d_2(\alpha^{2k})=0$. Since α is of dimension 1, $\alpha^2=Sq^1\alpha$. But α is transgressive; hence, so is α^2 and

$$d_3(\alpha^2)=\tau(\alpha^2)=\tau(Sq^1\alpha)=Sq^1(\tau\alpha)=Sq^1(\iota_2)$$

This shows that $E_2^{3,0}$ contains a Z_2 generated by $Sq^1(\iota_2)$. If we consider α^4, then, since $\alpha^4=(\alpha^2)^2=Sq^2Sq^1\alpha$, this element also is transgressive and there must be a Z_2 at (5,0) generated by $Sq^2Sq^1(\iota_2)$.

The reader may pursue this calculation further if he wishes. We remark that $H^5(B;Z_2)$ contains not only the Z_2 already mentioned but also a Z_2 generated by the product $(Sq^1\iota_2)(\iota_2)$ which arises because $d_2^{3,1}$ must be non-zero. Also, it is not hard to verify that α^n is transgressive if and only if n is a power of 2.

In fact, $H^*(Z_2,2;Z_2)$ is the polynomial ring over Z_2 generated by all $Sq^I(\iota_2)$ where I is an admissible sequence of excess less than 2. We can verify this in low dimensions by computations such as those outlined above; but to prove it in general we need better tools. A famous theorem of A. Borel is such a tool.

$H^*(Z_2,q;Z_2)$ AND BOREL'S THEOREM

A graded ring R over Z_2 is said to have the ordered set $x_1,x_2,\ldots$ as a *simple system of generators* if the monomials

$$\{x_{i_1}x_{i_2}\cdots x_{i_r}: i_1 < i_2 < \cdots < i_r\}$$

form a Z_2-basis for R and if for each n only finitely many x_i have gradation n.

Examples of rings with a simple system of generators include exterior algebras and the locally finite graded polynomial ring $Z_2[x_1,x_2,\ldots]$; in the latter case, the $\{x_i^{2^k}\}$ form a simple system of generators.

Theorem 1 (A. Borel)

Let $(E,p,B;F)$ be a fibre space with E acyclic, and suppose $H^*(F;Z_2)$ has a simple system $\{x_\alpha\}$ of transgressive generators. Then $H^*(B;Z_2)$ is the polynomial ring in the $\{\tau(x_\alpha)\}$.

A proof of this theorem is outlined in an appendix to this chapter.

Borel's theorem is precisely what we need in order to calculate $H^*(Z_2,2;Z_2)$; in fact, we have the following more general result.

Theorem 2

$H^*(Z_2,q;Z_2)$ is the polynomial ring over Z_2 with generators $\{Sq^I(\iota_q)\}$ where I runs through all admissible sequences of excess less than q.

The following proof of this theorem consists of an application of Borel's theorem and a lengthy exercise in admissible sequences.

We introduce the notation $L(p,r)$ for the sequence $2^{r-1}p,2^{r-2}p,\ldots,$ $4p,2p,p$ where $p>0$ and $r\geq 0$. (If $r=0$, we write $L(\ ,0)$.) Then the excess of $L(p,r)$ is p and the length of $L(p,r)$ is r; the degree is seen to be $p(2^r-1)$.

The theorem is known for $q=1$ (Proposition 2 of Chapter 2). We prove it by induction on q. We suppose it true for q and consider the fibre space

$$\begin{array}{rl} F=K(Z_2,q)\rightarrow & L \\ & \downarrow \\ & B=K(Z_2,q+1) \end{array}$$

By hypothesis $H^*(F;Z_2)$ is the polynomial ring over Z_2 with generators $\{Sq^I(\iota_q): I \text{ admissible}, e(I)<q\}$. To keep the notation intelligible, we write $H^*(F;Z_2)=P[\{z_j\}]$ where $\{z_j\}$ are the $Sq^I(\iota_q)$ suitably indexed. Let p_j denote the dimension of z_j (which is q plus the degree of the corresponding I). Then $(z_j)^{2^r}=Sq^{L(p_j,r)}(z_j)$. And the $\{(z_j)^{2^r}\}$ form a simple system of generators for $H^*(F;Z_2)$. Since ι_q is obviously transgressive, $\tau(\iota_q)=\iota_{q+1}$, all these generators are transgressive and Borel's theorem applies:

$$\begin{aligned} H^*(B;Z_2)&=P[\{\tau(z^{2^r})\}]=P[\{\tau Sq^{L(p_j,r)}z_j\}] \\ &=P[\{Sq^{L(p_j,r)}\tau(z_j)\}]=P[\{Sq^{L(p_j,r)}Sq^I\iota_{q+1}\}] \qquad (I \text{ from } z_j) \end{aligned}$$

Thus the theorem will be proved when we have shown that, as I runs through all the admissible sequences with $e(I)<q$, $L(q+d(I),\ r)\cdot I$ runs exactly once through all the admissible sequences with $e(I)<q+1$.

It is clear that every $L(\cdots)\cdot I$ is admissible. We will now construct an inverse function $J\rightarrow LI$ where J runs through the admissible sequences with $e(J)<q+1$; this will complete the proof.

Any admissible sequence $J=\{j_1,j_2,\ldots,j_s\}$ may be written (in at least one way) as $J=\{j_1,\ldots,j_t\}\cdot\{j_{t+1},\ldots,j_s\}$ where $j_i=2(j_{i+1})$ for all $i\leq t-1$. (Here t may be zero.) Then $J=L(j_t,t)\cdot I$ and $e(J)=j_t+e(I)-2(j_{t+1})$ if $t\geq 1$; $e(J)=e(I)$ if $t=0$. Recall the formula $e(I)=2i_1-d(I)$. Substituting this gives $e(J)=j_t-d(I)$ for $t\geq 1$.

Now, if $t=0$, each J with $e(J)<q$ has a unique expression $L(\ ,0)\cdot I$ with $e(I)<q$; if $e(J)=q$, then J obviously has a unique expression of the form LI.

If $t\geq 1$, we have $e(J)=j_t-d(I)$; hence $e(J)=q$ if and only if $j_t=q+d(I)$. Thus, if $e(J)<q$, J has a unique expression $L(\ ,0)\cdot I$ with $e(I)<q$; if $e(J)=q$, we choose t the maximum of all allowable t such that $J=L(q+d(I),\ t)\cdot I$ (with $e(I)<q$), namely, the t for which $j_t>2(j_{t+1})$.

This completes the proof of the theorem.

FURTHER SPECIAL CASES OF $H^*(\pi,n;G)$

The above results and methods enable us to assemble the following results about the cohomology of $K(\pi,n)$ with Z_2 coefficients.

Proposition 3

$H^*(Z,2;Z_2)$ is the polynomial ring over Z_2 generated by a two-dimensional class ι_2.

PROOF: Essentially the same as for $H^*(Z,2;Z)$ given earlier (Proposition 2).

Theorem 3

$H^*(Z,q;Z_2)$ is the polynomial ring with generators $\{Sq^I(\iota_q)\}$ where I runs through admissible sequences of excess $e(I)<q$ and where i_r, the last entry in I, is different from 1.

PROOF: By the same methods as the proof of Theorem 2. The proviso $i_r>1$ arises because $Sq^1(\iota_q)\doteq 0$.

Proposition 4

$H^*(Z_{2^m},1;Z_2)=P[d_m(\iota_1)]\otimes E(\iota_1)$ for $m\geq 2$. Here d_m denotes the mth Bockstein homomorphism of Chapter 7; $P[d_m(\iota_1)]$ denotes the polynomial ring over Z_2 on one generator $d_m(\iota_1)$; and $E(\iota_1)$, the exterior algebra on one generator ι_1.

PROOF: Compare the calculation for $m=1$ given earlier in this chapter.

Theorem 4

$H^*(Z_{2^m},q;Z_2)$ is the polynomial ring with generators $\{Sq^{I_m}(\iota_q)\}$ where we define $Sq^{I_m}=Sq^I$ if I terminates in $i_r>1$ and $Sq^{I_m}=Sq^{i_1}\cdots Sq^{i_{r-1}}d_m$ (i.e., Sq^I with d_m replacing Sq^{i_r}) if $i_r=1$; and where I runs through admissible sequences of excess $e(I)<q$.

Proof by the same methods as before.

DISCUSSION

The main results (Theorems 2, 3, and 4) of this chapter were proved by Serre. Theorem 2 completes at last the proof of Theorem 1 of Chapter 3.

In Chapter 1 we commented that we would study cohomology operations both by construction of explicit operations and by general study of $H^*(\pi,n;G)$. The reader should note that in the cases covered by Theorems 2, 3, and 4 the two approaches have merged. We have both classified the operations and specifically identified them, as well as having made in Chapters 3 and 6 detailed analysis of many of their properties.

Theorem 1A of the Appendix, a spectral-sequence comparison theorem, is by no means stated in full generality and is in fact a special case of a theorem of Zeeman.

We have used (and will use again) the following consequence of a result of Milnor. Let $f\colon X \to Y$ a map of spaces having the homotopy type of CW-complexes with a finite number of cells in each dimension. Convert f into a fibre map (by Proposition 1). Then the fibre is again such a space.

APPENDIX: PROOF OF BOREL'S THEOREM

We outline a proof of Borel's theorem based on a simplified version of the spectral-sequence comparison theorem.

In this appendix we consider only spectral sequences satisfying the following conditions:

1. $E_r^{p,q} = 0$ if $p < 0$ or $q < 0$
2. $E_2^{p,q} = E_2^{p,0} \otimes E_2^{0,q}$
3. $d_r^{p,q}$ has bidegree $(r, -r+1)$

Note that the Serre spectral sequence for the mod 2 cohomology of a fibre space has this form.

By a homomorphism $f\colon E \to \bar{E}$ of spectral sequences we mean a system of maps $\{f_r^{p,q}\}, f_r^{p,q}\colon E_r^{p,q} \to \bar{E}_r^{p,q}$ such that $\bar{d}_r f_r = f_r d_r, f_{r+1}$ is induced by f_r and $f_2^{p,q} = f_2^{p,0} \otimes f_2^{0,q}$. We now state a comparison theorem.

Theorem 1A

Let $f\colon E \to \bar{E}$ be a homomorphism of spectral sequences satisfying conditions (1) to (3). Suppose $f_2^{0,q}\colon E_2^{0,q} \to \bar{E}_2^{0,q}$ is an isomorphism for each q. Suppose $E_\infty^{p,q} = \bar{E}_\infty^{p,q} = 0$ except for $(p,q) = (0,0)$. Then $f_r^{p,q}$ is an isomorphism for all (p,q) and all $r \geq 2$.

REMARK: Under the hypotheses, the conclusion is equivalent to the assertion that $f_2^{p,0}$ is an isomorphism for each p.

The proof is by induction on the column p. By hypothesis, $f_2^{0,*}$ is an isomorphism. We indicate the induction step. Suppose we have proved $f_2^{s,0}$ an isomorphism for $s < p$. For $r < p$, $E_r^{p,0} = E_\infty^{p,0} = 0 = \bar{E}_\infty^{p,0} = \bar{E}_r^{p,0}$ and $f_r^{p,0}$ is an isomorphism. Now proceed by descent on r—i.e., assume $f_{r+1}^{p,0}$ is an isomorphism. Using the fact that $E_\infty^{p-r,r-1} = \bar{E}_\infty^{p-r,r-1} = 0$ and the induction hypothesis on p, it is easy to see that $f_r^{p,0}$ maps im $d_r^{p-r,r-1}$ isomorphically onto im $\bar{d}_r^{p-r,r-1}$. The five lemma and the diagram below

$$\begin{array}{ccccccc} 0 \to \operatorname{im} d_r^{p-r,r-1} & \to & E_r^{p,0} & \to & E_{r+1}^{p,0} & \to & 0 \\ \downarrow f_r^{p,0} & & \downarrow f_r^{p,0} & & \downarrow f_{r+1}^{p,0} & & \\ 0 \to \operatorname{im} \bar{d}_r^{p-r,r-1} & \to & \bar{E}_r^{p,0} & \to & \bar{E}_{r+1}^{p,0} & \to & 0 \end{array}$$

show that $f_r^{p,0}$ is an isomorphism. The downward induction on r then shows that $f_2^{p,0}$ is an isomorphism, which completes the induction on p and hence the proof.

We now apply Theorem 1A to prove Theorem 1. Take for $\bar{E}$ the spectral sequence in question, namely, the Serre sequence for an acyclic total space with $H^*(F;Z_2)$ having the simple system of transgressive generators $\{x_\alpha\}$. For ease, assume deg $x_\alpha \geq 1$. We will construct a spectral sequence E and a map f: $E \to \bar{E}$ in such a way that the comparison theorem applies.

Let P be the graded polynomial ring over Z_2 with one generator y_α for each x_α and with deg $y_\alpha =$ deg $x_\alpha + 1$. Let Λ be the graded exterior algebra over Z_2 with one generator z_α for each x_α and with deg $z_\alpha =$ deg x_α. Let $R = P \otimes \Lambda$. Define a differential δ of total degree 1 on R by the formula $\delta(y \otimes z_\alpha) = yy_\alpha$ for $y \in P$ and the requirement that δ be a derivation.

Filter R by setting $F^p(R) = \sum_{i \geq p} P^i \otimes \Lambda$ and consider the resulting spectral sequence E for $H^*(R)$. It is easy to check that $E_2^{p,q} = R^{p,q}$, all (p,q). Thus E satisfies properties (1) to (3). The acyclicity is easily verified if $\{x_\alpha\}$ has only one element; the general case follows by the Künneth formula.

Define a ring homomorphism $f_2^{*,0}$: $P \to H^*(B;Z_2)$ by $y_\alpha \to \tau(x_\alpha)$ and an additive (not, in general, multiplicative) homomorphism $f_2^{0,*}$: $\Lambda \to H^*(F;Z_2)$ by $z_{i_1} \cdots z_{i_r} \to x_{i_1} \cdots x_{i_r}$, $i_1 < \cdots < i_r$. Define $f_2^{p,q} = f_2^{p,0} \otimes f_2^{0,q}$. Once f_r is defined, define f_{r+1} by requiring that f_{r+1} be induced by f_r. This is possible since $f_r d_r = \bar{d}_r f_r$, which is readily checked (and uses the hypothesis that each x_α is transgressive).

We now may apply Theorem 1A. It follows that $f_2^{*,0}$: $P \to H^*(B;Z_2)$ is a ring isomorphism.

REFERENCES

General

1. H. Cartan [2,5].
2. A. Dold [1].
3. J.-P. Serre [2].
4. N. E. Steenrod [2].

Milnor's theorem

1. J. Milnor [3].

The spectral-sequence comparison theorem

1. E. C. Zeeman [1].

CHAPTER 10

CLASSES OF ABELIAN GROUPS

In subsequent chapters we will use the Steenrod operations to make many calculations. The material of this chapter will be an important tool in the calculations and is of interest in its own right.

Let A be an abelian group. By the *p-component* of A, where p is a prime integer, we mean the quotient group obtained from A by factoring out the subgroup of all torsion elements of order prime to p. For example, if $A = Z \oplus Z_2 \oplus Z_3$, then the 2-component of A is $A/Z_3 \approx Z \oplus Z_2$.

In this chapter we will obtain the following result: Suppose X is a simply connected space such that $H_k(X)$ is finitely generated for all k. Suppose Y is another such space and that we have a map $f: X \to Y$ such that f induces an isomorphism on cohomology with Z_p coefficients. Then the p-components of the corresponding homotopy groups are isomorphic.

This result will be a fundamental tool in our calculations of homotopy groups. Its proof is based on Serre's theory of classes of abelian groups.

ELEMENTARY PROPERTIES OF CLASSES

Definition

A *class of abelian groups* is a collection $\mathcal{C}$ of abelian groups satisfying the following axiom:

1. If $0 \to A' \to A \to A'' \to 0$ is a short exact sequence, then A is in $\mathcal{C}$ if and only if both A' and A'' are in $\mathcal{C}$

Thus a class of abelian groups is closed under the formation of subgroups, quotient groups, and group extensions.

When we use the letter $\mathcal{C}$, it will be implicit that $\mathcal{C}$ is a non-empty class of abelian groups.

Definition

A homomorphism $f: A \to B$ is said to be a $\mathcal{C}$-*monomorphism* if $\ker f \in \mathcal{C}$, a $\mathcal{C}$-*epimorphism* if $\operatorname{coker} f \in \mathcal{C}$, and a $\mathcal{C}$-*isomorphism* if both $\ker f$ and $\operatorname{coker} f$ are in $\mathcal{C}$.

WARNING: a $\mathcal{C}$-isomorphism may fail to have an inverse.

Consider the relation $\sim$ defined by $A \sim B$ if and only if there exists a $\mathcal{C}$-isomorphism of A to B. This relation is reflexive but need not be symmetric. We say that two groups A,B are $\mathcal{C}$-*isomorphic* if they are equivalent by the smallest reflexive, symmetric, and transitive relation containing the above relation. Thus A and B are $\mathcal{C}$-isomorphic if and only if there exists a finite sequence $\{A = A_0, A_1, A_2, \ldots, A_n = B\}$ and, for each i $(0 \leq i < n)$, a $\mathcal{C}$-isomorphism between A_i and A_{i+1} in one direction or the other.

We will have need of the following further axioms at various times.

2A. If $A,B \in \mathcal{C}$, then $A \otimes B \in \mathcal{C}$ and $\operatorname{Tor}(A,B) \in \mathcal{C}$
2B. If $A \in \mathcal{C}$, then $A \otimes B \in \mathcal{C}$ for every abelian group B
3. If $A \in \mathcal{C}$, then $H_n(A,1;Z) \in \mathcal{C}$ for every $n > 0$

Note that (2B) implies (2A), for $\operatorname{Tor}(A,B)$ is a subgroup of $A \otimes R$ for a certain group R.

The most important examples in our work are the following.

Example 1

$\mathcal{C}_0$, the trivial class, containing only one group, the trivial group. This class clearly satisfies Axioms (1), (2B), and (3).

Example 2

$\mathcal{C}_{FG}$, the class of finitely generated (abelian) groups. This class satisfies Axioms (1), (2A), and (3). A counterexample for Axiom (2B) is given by $A =$ the integers, $B =$ the rationals. As for the proof of (3), it is enough to verify (3) for Z and for Z_p, p prime. It is clear for Z. For Z_2, we have a cell structure of finite type, from which the result is clear. An analogous cell structure can be given for $p > 2$. We omit the details.

Example 3

$\mathcal{C}_p$, where p is a prime: the class of abelian torsion groups of finite exponent such that the order of every element is prime to p. This class satisfies Axioms (1), (2B), and (3). The proof of (3) is not obvious and we will give it later.

If the class $\mathcal{C}$ in question is taken to be $\mathcal{C}_0$ (Example 1), it is readily seen that all the $\mathcal{C}$-notions, such as $\mathcal{C}$-isomorphism, reduce to the usual classical notions.

We remark that many standard results of homological algebra have analogues in $\mathcal{C}$-theory; for example, there is a "five-lemma mod $\mathcal{C}$."

TOPOLOGICAL THEOREMS MOD $\mathcal{C}$

We are interested in the following $\mathcal{C}$-theory generalizations of three classical topological theorems which have been discussed in earlier chapters.

Theorem 1 (Hurewicz theorem mod $\mathcal{C}$)

Let X be a 1-connected space. Let $\mathcal{C}$ be a class satisfying Axioms (1), (2A), and (3). If $\pi_i(X) \in \mathcal{C}$ for all $i < n$, then $H_i(X) \in \mathcal{C}$ for all $i < n$ and the Hurewicz homomorphism $h: \pi_n(X) \to H_n(X)$ is a $\mathcal{C}$-isomorphism.

Theorem 2 (Relative Hurewicz theorem mod $\mathcal{C}$)

Let A be a subspace of X; let X and A be 1-connected, and suppose also that $i_\# : \pi_2(A) \to \pi_2(X)$ is an epimorphism. Let $\mathcal{C}$ be a class satisfying Axioms (1), (2B), and (3). Then if $\pi_i(X,A) \in \mathcal{C}$ for all $i < n$, it follows that $H_i(X,A) \in \mathcal{C}$ for all $i < n$ and $h: \pi_n(X,A) \to H_n(X,A)$ is a $\mathcal{C}$-isomorphism.

Theorem 3 (Whitehead's theorem mod $\mathcal{C}$)

Let f be a mapping $A \to X$, where A and X are 1-connected, and suppose $f_\#$ is an isomorphism on π_2. Let $\mathcal{C}$ satisfy (1), (2B), and (3). Then $f_\#$ is a $\mathcal{C}$-isomorphism on π_i for all $i < n$ and a $\mathcal{C}$-epimorphism on π_n if and only if the corresponding statements hold for f_* on H_*.

From the Hurewicz theorem mod $\mathcal{C}_{FG}$ (see Example 2) we can immediately draw a useful consequence.

Proposition 1

All homotopy groups of a finite 1-connected complex are finitely generated.

The proof is immediate from the Hurewicz theorem, in the light of the following obvious property of classes.

Proposition 2

If the groups A,B are $\mathcal{C}$-isomorphic and $A \in \mathcal{C}$, then $B \in \mathcal{C}$. (Here $\mathcal{C}$ need only satisfy Axiom (1).)

We will indicate the proofs of the above theorems. First we make some remarks concerning spectral sequences.

Remark 1

If $E^2_{p,q} \in \mathcal{C}$, then $E^r_{p,q} \in \mathcal{C}$ for all $r \geq 2$.

This is clear from Axiom (1).

Remark 2A

If $\mathcal{C}$ satisfies (2A), then, in the spectral sequence of a fibre space, if $E^2_{p,0} \in \mathcal{C}$, $0 \leq p \leq m$, and $E^2_{0,q} \in \mathcal{C}$, $0 \leq q \leq n$, it follows that $E^2_{p,q} \in \mathcal{C}$ for all p,q in the given range.

Remark 2B

If $\mathcal{C}$ satisfies (2B) and $H_n(B) \in \mathcal{C}$, where B is the base space of a fibre space, then $E^2_{n,q} \in \mathcal{C}$, for all q, in the spectral sequence.

The proofs are easy.

THE HUREWICZ THEOREM

We now indicate the proof of the Hurewicz theorem.

Recall (Theorem 3 of Chapter 8) that we proved this theorem for the case $\mathcal{C} = \mathcal{C}_0$ by spectral-sequence methods, using induction on n. The induction began by using classical methods to prove the anomalous case $n = 1$. The essential step in the proof is the sequence of isomorphisms

$$\pi_n(X) \approx \pi_{n-1}(\Omega X) \approx H_{n-1}(\Omega X) \approx H_n(X)$$

where the first isomorphism is well known, the second comes out of the induction hypothesis, and the third is proved by means of the spectral sequence of the fibre space $(E,p,X;\ \Omega X)$ where E is contractible.

In the present case, we assume X is 1-connected, and so the induction begins without trouble, but we cannot apply the induction hypothesis to ΩX because it may fail to be simply connected. We are therefore obliged to introduce T, the universal cover of ΩX, and use the following detour to obtain what corresponds to the second isomorphism in the above sequence:

$$\pi_{n-1}(\Omega X) \approx \pi_{n-1}(T) \underset{\mathcal{C}}{\approx} H_{n-1}(T) \underset{\mathcal{C}}{\approx} H_{n-1}(\Omega X)$$

where we write $A \underset{\mathcal{C}}{\approx} B$ or, sometimes, $A \approx B \pmod{\mathcal{C}}$ to indicate that A and B are $\mathcal{C}$-isomorphic. Since X is implicitly assumed "nice," the theorem of Milnor (see Chapter 9) permits us to assume that ΩX (up to homotopy type) has a universal cover T.

Thus our proof consists of establishing five ($\mathcal{C}$-)isomorphisms. The first of these, $\pi_n(X) \approx \pi_{n-1}(\Omega X)$, is well known (Chapter 8, Example 3).

It is also well known that $\pi_{n-1}(\Omega X) \approx \pi_{n-1}(T)$. Indeed, the universal cover of a space has the same homotopy groups as the space except that it is simply connected; but, as our theorem is obvious for $n=2$, we may assume in our proof that $n>2$; hence the result.

The next step is the $\mathcal{C}$-isomorphism of $\pi_{n-1}(T)$ and $H_{n-1}(T)$. But $\pi_i(T) \approx \pi_i(\Omega X) \approx \pi_{i+1}(X)$ for $i>1$; thus T satisfies the hypotheses of the theorem for dimension $n-1$, and thus the result is a consequence of the induction hypothesis.

The next, and crucial, step is to establish that $H_i(T) \approx H_i(\Omega X)$ (mod $\mathcal{C}$) for all $i<n$. This depends on arguments in the spectral sequence of the fibre space $(\Omega X, p, K(\pi,1))$, with fibre T, where we write π for $\pi_1(\Omega X) \approx \pi_2(X)$. Since the base space is not simply connected, it is not immediately clear that the spectral sequence of Chapter 8 is applicable; in fact, it does apply. We omit the proof. The construction of the fibre space itself may be carried out as follows: Let W be a π-free acyclic complex. Since π acts on T, we may define $W \times_\pi T$ in the customary way as the quotient of $W \times T$ by the relation $(wg,t) \sim (w,gt)$ $(g \in \pi)$. This is the total space; since W is acyclic, $W \times_\pi T \simeq (\text{pt.}) \times_\pi T \simeq T/\pi \simeq \Omega X$. The total space maps naturally to W/π, that is, to $K(\pi,1)$, which is the base space. (Compare the construction of $K(Z_2,1)$ in Chapter 2.) The fibre is clearly T.

Consider, then, the spectral sequence of this fibre space. Since $\pi=\pi_2(X)$ is in $\mathcal{C}$, $E^2_{p,0} \approx H_p(\pi,1;Z)$ is in $\mathcal{C}$, for all p, by Axiom (3). By the induction hypothesis, $E^2_{0,q} \approx H_q(T) \in \mathcal{C}$ for all $q<n-1$. Using Remark 2A, we have $E^2_{p,q} \in \mathcal{C}$ for all p, if $q<n-1$. Thus clearly $H_i(\Omega X) \in \mathcal{C}$ for $i<n-1$. We now must show that $H_{n-1}(\Omega X)$ and $H_{n-1}(T)$ are $\mathcal{C}$-isomorphic. Clearly $H_{n-1}(\Omega X)$ is $\mathcal{C}$-isomorphic to $E^\infty_{0,n-1}$ and $H_{n-1}(T) \approx E^2_{0,n-1}$. But $E^2_{0,n-1}$ and $E^\infty_{0,n-1}$ are $\mathcal{C}$-isomorphic, for $E^3_{0,n-1}$ is the quotient of $E^2_{0,n-1}$ by a group in $\mathcal{C}$; similarly for $E^4_{0,n-1}$, etc.

This establishes the isomorphism $H_i(T) \approx H_i(\Omega X)$ (mod $\mathcal{C}$) for all $i<n$.

Finally we must show $H_{n-1}(\Omega X) \approx H_n(X)$ (mod $\mathcal{C}$). Here the argument is nothing but the $\mathcal{C}$-theory form of the corresponding argument in the case $\mathcal{C}=\mathcal{C}_0$. We use the spectral sequence of the contractible fibre space (E,p,X) with fibre ΩX. We have shown above that $E^2_{0,q} \approx H_q(\Omega X) \in \mathcal{C}$ for $q<n$. Also, by the induction hypothesis, $E^2_{p,0} \approx H_p(X) \approx \pi_p(X) \in \mathcal{C}$ for $p<n$. Now $H_n(X) \approx E^2_{n,0}$. As argued above, $E^2_{n,0} \approx E^n_{n,0}$ (mod $\mathcal{C}$). But E is contractible, and so $d^n\colon E^n_{n,0} \to E^n_{0,n-1}$ is a $\mathcal{C}$-isomorphism. Again, $E^n_{0,n-1} \approx E^2_{0,n-1}$ (mod $\mathcal{C}$), but this is just $H_{n-1}(\Omega X)$. Thus $H_n(X) \approx H_{n-1}(\Omega X)$ (mod $\mathcal{C}$).

This gives us the five isomorphisms which yield $\pi_n(X) \approx H_n(X)$ (mod $\mathcal{C}$) and hence the Hurewicz theorem.

THE RELATIVE HUREWICZ THEOREM

In order to prove the relative Hurewicz theorem, we introduce the *relative fibre space*

$$\begin{array}{c} \Omega X \longrightarrow (E, Y) \\ \downarrow \\ (X, A) \end{array}$$

where $(E,p,X;\ \Omega X)$ is the usual contractible fibre space over X and $Y = p^{-1}(A)$. It gives rise to a spectral sequence, entirely analogous to that for a fibre space, converging to $H_*(E, Y)$ and with $E^2_{p,q} = H_p(X,A;\ H_q(\Omega X))$.

Recall that X and A are assumed 1-connected and that the hypotheses of the theorem also imply $\pi_2(X,A) = 0$. We operate by induction on n; the theorem is obvious for $n = 2$; we assume it for $n - 1$ and deduce it for n. It is immediate from the induction hypothesis that $H_i(X,A) \in \mathcal{C}$ for $i < n$. We must show $H_n(X,A) \approx \pi_n(X,A)$ (mod $\mathcal{C}$).

In the spectral sequence of the relative fibre space, since $H_i(X,A) \in \mathcal{C}$ for $i < n$, we have $E^2_{p,q} \in \mathcal{C}$ for $p < n$ (for all q), using Remark 2B. By arguments similar to those of the proof of the Hurewicz theorem,

$$H_n(E, Y) \underset{\mathcal{C}}{\approx} E^\infty_{n,0} \underset{\mathcal{C}}{\approx} E^2_{n,0} \approx H_n(X,A)$$

On the other hand, $\pi_j(X,A) \approx \pi_j(E, Y)$ by the covering homotopy property and $\pi_j(E, Y) \approx \pi_{j-1}(Y)$ because E is contractible. Putting $j = 2$ shows that Y is simply connected and the Hurewicz theorem applies to Y, giving $\pi_{n-1}(Y) \approx H_{n-1}(Y)$ (mod $\mathcal{C}$). Thus

$$\pi_n(X,A) \approx \pi_n(E, Y) \approx \pi_{n-1}(Y) \underset{\mathcal{C}}{\approx} H_{n-1}(Y) \approx H_n(E, Y) \underset{\mathcal{C}}{\approx} H_n(X,A)$$

which proves the theorem.

The reader should now be able to deduce the Whitehead theorem from the relative Hurewicz theorem by essentially the method given in Chapter 8.

$\mathcal{C}_p$ SATISFIES AXIOM 3

We now focus our attention on the class $\mathcal{C}_p$. Recall that a group G belongs to $\mathcal{C}_p$ if $nG = 0$ for some integer n and if every $g \in G$ has finite order prime to p.

It is clear that $\mathcal{C}_p$ satisfies Axioms (1) and (2B). We now prove that Axiom (3) also holds. Thus let $A \in \mathcal{C}_p$; we show that $H_n(A,1;Z) \in \mathcal{C}_p$ for all $n \geq 1$.

Indeed, let A have exponent q prime to p. We prove by induction on n that there exist non-negative integers e_n such that $H_n(A,1)$ has exponent dividing q^{e_n}, $n \geq 1$. This assertion is trivial for $n=1$.

Let s be a positive integer. Multiplication by s in A induces on $K(A,1)$ the map $s: K(A,1) \to K(A,1)$ given by

$$K(A,1) \xrightarrow{\Delta} K_1(A,1) \times \cdots \times K_s(A,1) \xrightarrow{\mu} K(A,1)$$

where Δ is the diagonal and μ the iterated multiplication. Note that $q: K(A,1) \to K(A,1)$ is null-homotopic.

Consider $H_n(\prod_{i=1}^{s} K_i(A,1))$ as given by the Künneth formula. If we use the induction hypothesis and Axiom (2A) on $\mathcal{C}_p$, it follows that this group may be written as $\sum_{i=1}^{s} H_n(K_i(A,1)) \oplus G$ where G has exponent dividing q^f for some f.

Let $a \in H_n(K(A,1))$. Then we may write $\Delta_*(a) = \sum_{i=1}^{s} a_i \oplus g$ for some $g \in G$. Thus $s_*(a) = sa + \mu_*(g)$. Taking $s=q$, recalling $q_*(a)=0$, we have $qa + \mu_*(g) = 0$ for some $g \in G$. Thus $q^{f+1}a = 0$ and we may take $e_n = f+1$. This completes the proof.

We now state for future use another result about $\mathcal{C}_p$.

Lemma 1

Let $f: A_1 \to A_2$ be a homomorphism of finitely generated abelian groups. Suppose f is a $\mathcal{C}_p$-isomorphism. Then A_1 and A_2 have isomorphic p-components.

PROOF: Let $p(A_i)$ denote the subgroup of A_i consisting of elements of finite order prime to p. We must show $A_1/p(A_1) \approx A_2/p(A_2)$. Clearly $f[p(A_1)] \subset p(A_2)$. Since $A_i \in \mathcal{C}_{FG}$, $p(A_i) \in \mathcal{C}_p$. Thus we have the diagram

$$\begin{array}{ccccccccc} 0 \to & p(A_1) & \to & A_1 & \to & A_1/p(A_1) & \to 0 \\ & \downarrow & & \downarrow f & & \downarrow \hat{f} & \\ 0 \to & p(A_2) & \to & A_2 & \to & A_2/p(A_2) & \to 0 \end{array}$$

By the 5-lemma mod $\mathcal{C}_p$, $\hat{f}$ is a $\mathcal{C}_p$-isomorphism. Using this fact, the structure theorem for finitely generated abelian groups, and the observation that $A_i/p(A_i)$ has no torsion prime to p, the reader may easily deduce the following: $\hat{f}$ is a monomorphism, induces an isomorphism of torsion subgroups, and has for its image a subgroup of maximum rank. The result follows.

We remark that $\hat{f}$ need not be an isomorphism.

THE $\mathcal{C}_p$ APPROXIMATION THEOREM

We can now give the theorem announced in the introduction to this chapter.

Theorem 4 ($\mathcal{C}_p$ Approximation Theorem)

Let X and A be simply connected (nice) spaces such that $H_i(A)$ and $H_i(X)$ are finitely generated for every i. Let f be a map $A \to X$ such that $f_\# : \pi_2(A) \to \pi_2(X)$ is epimorphic. (We may assume, without loss of generality, that f is an inclusion.) Then conditions (1) to (6) below are equivalent and imply condition (7).

1. $f^*: H^i(X;Z_p) \to H^i(A;Z_p)$ is isomorphic for $i < n$ and monomorphic for $i = n$
2. $f_*: H_i(A;Z_p) \to H_i(X;Z_p)$ is isomorphic for $i < n$ and epimorphic for $i = n$
3. $H_i(X,A; Z_p) = 0$ for $i \leq n$
4. $H_i(X,A; Z) \in \mathcal{C}_p$ for $i \leq n$
5. $\pi_i(X,A) \in \mathcal{C}_p$ for $i \leq n$
6. $f_\# : \pi_i(A) \to \pi_i(X)$ is $\mathcal{C}_p$-isomorphic for $i < n$ and $\mathcal{C}_p$-epimorphic for $i = n$
7. $\pi_i(A)$ and $\pi_i(X)$ have isomorphic p-components for $i < n$

Thus this theorem reduces the problem of computing the p-component of $\pi_i(X)$ to that of finding a space A with the same cohomology in Z_p coefficients, together with a map of A into X inducing isomorphisms in Z_p cohomology.

The proof is not difficult with the tools we now have at hand.

Conditions (1) and (2) are equivalent by vector-space duality, since Z_p is a field and all H_i are finitely generated.

Conditions (2) and (3) are equivalent by the exact homology sequence of the pair (X,A).

To see that (3) implies (4), recall that we have, from the universal coefficient theorems, an exact sequence

$$0 \to H_i(X,A) \otimes Z_p \to H_i(X,A; Z_p) \to \operatorname{Tor}(H_{i-1}(X,A), Z_p) \to 0$$

From (3), the group in the middle is zero, and so the group on the left is also zero. But $H_i(X,A)$ is finitely generated, so that $H_i(X,A) \otimes Z_p = 0$ implies that $H_i(X,A)$ must be the direct sum of finite groups of order prime to p; thus $H_i(X,A) \in \mathcal{C}_p$, which is condition (4).

Conversely, from (4) (applied both to H_i and H_{i-1}) and the above exact sequence, we have (3), by Axiom (2B).

Conditions (4) and (5) are equivalent by the relative Hurewicz theorem.

Conditions (5) and (6) are equivalent by the exact homotopy sequence, mod $\mathcal{C}_p$, of the pair (X,A).

Now by the Hurewicz theorem mod $\mathcal{C}_{FG}$, the groups $\pi_i(A)$ and $\pi_i(X)$ are finitely generated. Thus by Lemma 1, (6) implies (7).

This completes the proof.

DISCUSSION

Of the results of this chapter, we shall make the most use of Theorem 4, the $\mathcal{C}_p$ approximation theorem. It is this theorem which will justify our calculations of 2-components of some homotopy groups of spheres (and which justifies similar computations in the p-component for odd primes p).

The material presented here was developed by Serre. The reader may refer to his paper for a fuller discussion of relative fibre spaces.

Rather than show directly the triviality of the coefficient system in the spectral sequence for the fibre space $(\Omega X,p,K(\pi,1))$ with fibre T, we remark that in this particular case the existence of the spectral sequence was known before Serre's work.

EXERCISE

1. Prove Theorem 3.

REFERENCES

General
1. S.-T. Hu [1].
2. J.-P. Serre [1,3].
3. E. H. Spanier [1].

The spectral sequence of a covering
1. H. Cartan [4, Exp. XI–XIII].
2. — and S. Eilenberg [1].

CHAPTER 11

MORE ABOUT FIBRE SPACES

In the chapter following this we will launch our calculations of homotopy groups. The present chapter consists of several lemmas about fibre spaces which will be needed in the calculations and in the later development.

INDUCED FIBRE SPACES

Let ξ denote a fibre space $(E,p,B;\ F)$ and let f be a map of a space X into the base space B. The *induced fibre space* $f^*(\xi)$ is the fibre space with X as base space, F as fibre, and as total space the space $E' = \{(x,e): f(x) = p(e)\}$ topologized as a subspace of $X \times E$. The projection map of $f^*(\xi)$ is the natural projection $\pi_1 : E' \to X : (x,e) \to x$. Note that there is also a natural map $\pi_2 : E' \to E: (x,e) \to e$. It is easy to verify that if ξ is a fibre space in the sense of Serre or of Hurewicz, then $f^*(\xi)$ is likewise (see Exercise 1).

Observe that if X is a subspace of B and f is the inclusion, then E' may be identified in a natural way with a subspace of E, namely, $p^{-1}(X)$, and $f^*(\xi)$ in this case is just the restriction of ξ to the smaller base space X.

Two fibre spaces, $\xi = (E,p,B;\ F)$ and $\xi' = (E',p,B;\ F')$, are said to be *fibre homotopy equivalent* if there is a homotopy equivalence between the total spaces which is compatible with the projections. Precisely stated, this means that there exists a homotopy equivalence (φ,ψ), $\varphi : E \to E'$, $\psi : E' \to E$, where the homotopies $H: E \times I \to E$ and $H': E' \times I \to E'$, which deform the compositions $\psi\varphi$ and $\varphi\psi$ into the respective identity maps, are required to satisfy $p(H(t,e)) = p(e)$ and $p'(H'(t,e')) = p'(e')$ for all points e,e' and all $t \in I$. It is obvious that this gives an equivalence relation among fibre spaces over a fixed base space B and that the restriction of (φ,ψ) gives a homotopy equivalence of the fibres. In Exercise 2 the reader is asked to verify that homotopic maps induce fibre homotopy equivalent fibre spaces.

We will make extensive use of the next result, which shows the importance of the idea of induced fibre space.

Proposition 1

Suppose we have the situation shown in the diagram, where $\xi = (E,p,B)$

$$\begin{array}{ccccc} & & E' & \xrightarrow{\pi_2} & E \\ & {\scriptstyle h}\nearrow & \downarrow{\scriptstyle \pi_1} & & \downarrow{\scriptstyle p} \\ Y & \xrightarrow[g]{} & X & \xrightarrow[f]{} & B \end{array}$$

is a Hurewicz fibre space and E' is the induced fibre space $f^*(\xi)$. Suppose the composition fg: $Y \to B$ is null-homotopic. Then there exists a lifting of g, that is, a map h: $Y \to E'$ such that $\pi_1 h = g$. The conclusion also holds if ξ is only a fibre space in the sense of Serre, provided that Y is a finite complex.

PROOF: Using the covering homotopy property, we lift a null-homotopy of fg to obtain a map h_E: $Y \to E$ such that $ph_E = fg$. Then the maps g and h_E combine to give a map $h = g \times h_E$: $Y \to X \times E$ which has its image in E' and which obviously satisfies $\pi_1 h = g$.

It may seem trivial to remark that if the space E in the above proposition is a contractible space, then the converse holds—that is, $fg \simeq 0$ if and only if g may be lifted to E'. (This is obvious, since if g has a lifting h, then fg factors through E, using π_2.) However, this remark will have wide application, since an important case of induced fibre space is that in which ξ is the standard fibre space of paths over B (Chapter 8, Example 3).

THE TRANSGRESSION OF THE FUNDAMENTAL CLASS

Let $\xi = (E,p,B; F)$ be a fibre space such that the fibre F is $(n-1)$-connected. Then there is a fundamental class $\iota_F \in H^n(F;\pi_n(F))$. Moreover, the transgression of ι_F is defined, in the narrow sense. By definition, the *characteristic class* $\chi(\xi)$ of the fibre space ξ is this cohomology class $\tau(\iota_F) \in H^{n+1}(B; \pi_n(F))$.

Let E be the standard contractible fibre space over $K(\pi, n+1)$ (where π is an abelian group), that is, the fibre space of paths. Then the fibre is

$\Omega K(\pi, n+1) = K(\pi,n)$. From the cohomology spectral sequence of this fibre space, it is clear that the transgression

$$\tau\colon H^n(K(\pi,n);\pi) \to H^{n+1}(K(\pi, n+1);\pi)$$

is an isomorphism. Denote the fundamental classes of the base space and fibre by ι_{n+1} and ι_n, respectively. We claim that the above isomorphism maps ι_n to ι_{n+1}. This is certainly plausible; we urge the reader to verify it for himself, using the definition of fundamental class (Chapter 1) and the formulation of τ in terms of p^* and δ (see the discussion of transgression in Chapter 8).

Suppose now that f is a map of a space X into $K(\pi, n+1)$ and let $(E', q, X; K(\pi,n))$ be the induced fibre space. We then obtain a map of the Serre exact sequences in cohomology (the coefficients, which lie in π, are

$$\begin{array}{ccccc}
\cdots \to H^n(K(\pi,n)) & \xrightarrow{\tau_p} & H^{n+1}(K(\pi, n+1)) & \xrightarrow{p^*} & H^{n+1}(E) \\
\downarrow = & & \downarrow f^* & & \downarrow \pi_2^* \\
\cdots \to H^n(K(\pi,n)) & \xrightarrow[\tau_q]{} & H^{n+1}(X) & \xrightarrow[q^*]{} & H^{n+1}(E')
\end{array}$$

suppressed for clarity), where commutativity for the square containing the transgressions may be verified using the formulation in terms of p^* and δ. We therefore have the following relation,

Formula 1 $$f^*(\iota_{n+1}) = \tau_q(\iota_n)$$

which is fundamental in what follows. Since $f^*(\iota_{n+1})$ is just the cohomology class which represents f, in the sense of Chapter 1 (Theorem 1), the above relation may be stated thus: in the fibre space induced by f from the standard fibre space of paths, the fundamental class transgresses to the cohomology class which corresponds to f. Observe that $q^*(f^*(\iota_{n+1})) = 0$ in $H^{n+1}(E';\pi)$, by exactness.

BOCKSTEINS AND THE BOCKSTEIN LEMMA

Recall (from Chapter 7) the Bockstein exact couple,

$$\begin{array}{ccc}
H^*(\ ;Z) & \xrightarrow{(\times 2)} & H^*(\ ;Z) \\
\beta \nwarrow & & \swarrow \rho \\
& H^*(\ ;Z_2) &
\end{array}$$

where β denotes the Bockstein homomorphism of the exact sequence $0 \to Z \to Z \to Z_2 \to 0$ and ρ denotes reduction mod 2. In the resulting spectral sequence, d_r is the "rth Bockstein operator"; d_1 in particular is the homomorphism $\delta_2 = \rho\beta$ of Chapter 3.

The operator d_r has been defined in terms of the spectral sequence, so that its domain and range appear, for large values of r, as complicated "subquotients" of $H^*(\ ;Z_2)$. However, it is clear from the definition that the domain is a quotient group of the subgroup of $H^*(\ ;Z_2)$ of elements for which $d_i = 0$, $i < r$, or in other words a quotient group of $\bigcap_{i=1}^{r-1} \ker (d_i)$. Moreover, it is convenient to consider d_r as defined on $\bigcap_{i=1}^{r-1} \ker (d_i)$ itself, rather than on a quotient of this group. In this way we can make sense of a composition of Bockstein operators $d_r d_s$ even when $r \neq s$; and it is clear from the definitions that such a composition is always zero, modulo the indeterminacy of d_r. (When a homomorphism, such as d_r, takes values in a quotient group A/B, we say that d_r has *indeterminacy B*.)

Recall also from Chapter 7 that d_i vanishes on any cohomology class which is the image, under ρ, of a class in coefficients 2^{i+1}. Therefore the composite $d_i\rho$ gives a well-defined map $H^n(\ ;Z_{2^i}) \to H^{n+1}(\ ;Z_2)$, and $d_i\rho$ may be identified with a map (or rather a homotopy class of maps) from $K(Z_{2^i},n)$ into $K(Z_2, n+1)$.

In Chapter 9 we indicated a procedure for constructing a fibre space from a short exact sequence of abelian groups. Another technique which will be useful in the present context is the following.

Lemma 1

The map $d_i\rho\colon K(Z_{2^i},n) \to K(Z_2, n+1)$ defined above induces, from the standard fibre space of paths over $K(Z_2, n+1)$, a fibre space with fibre $K(Z_2,n)$ and total space $K(Z_{2^{i+1}},n)$. Moreover, the injection and projection of the induced fibre space correspond to the maps in the exact sequence

$$0 \to Z_2 \to Z_{2^{i+1}} \to Z_{2^i} \to 0$$

(under the correspondence of Chapter 1, Corollary 1).

The proof is easy, with the machinery available to us. The fibre is clearly $\Omega K(Z_2, n+1) = K(Z_2,n)$. From the homotopy exact sequence of the induced fibre space, it is clear that the total space must be $K(G,n)$ where $G/Z_{2^i} \approx Z_2$. To see that the group extension is the non-trivial one, write the spectral sequence (in Z_2 cohomology) of the induced fibre space and observe that $\tau(\iota_F) = (d_i\rho)^*(\iota_B)$ by Formula 1, where ι_F and ι_B denote the fundamental classes in the fibre and base of the induced fibre space. Thus we see from this spectral sequence that $H^n(E';Z_2) \approx Z_2$, so that G is cyclic; hence $G \approx Z_{2^{i+1}}$, as claimed. Finally, since the injection and projection are clearly not null-homotopic, they must represent the homomorphisms stated, since those are the only non-trivial ones between these groups.

Thus the above technique using the induced fibre space produces the same fibre space as the technique of Chapter 9. This is actually a special case of a much more general result about the classification of fibre spaces with $K(\pi,n)$ as fibre. The more general result will be obtained in Chapter 13.

The next lemma needed for our calculations is the following, known as the "Bockstein lemma."

Theorem 1

Let $(E,p,B;F)$ be a fibre space. Let the class $u \in H^n(F;Z_2)$ be transgressive, and suppose that, for some integer i $(i \geq 1)$ and for some class $v \in H^n(B;Z_2)$, $d_i v = \tau(u)$. Then $d_{i+1}p^*v$ is defined, and moreover

$$j^* d_{i+1} p^*(v) = d_1(u)$$

where j is the inclusion $F \subset E$. Here the members of the formula $d_i v = \tau(u)$ and of the formula of the conclusion lie in appropriate quotient groups of $H^{n+1}(F;Z_2)$. The equalities should be suitably interpreted.

This theorem is proved by the "method of universal example": we first prove it in a special case, or "universal example," and then show that the general case can be mapped into the special case in such a way that the general conclusion follows by naturality.

For the universal example, take the fibre space

$$\begin{array}{ccc} F = K(Z_2,n) & \xrightarrow{j} & E = K(Z_{2^{i+1}},n) \\ & & \downarrow \\ & & B = K(Z_{2^i},n) \end{array}$$

which was discussed in Lemma 1. For u, take the fundamental class $\iota_F \in H^n(F;Z_2)$, and for v, take $\rho(\iota_B) \in H^n(B;Z_2)$, the reduction mod 2 of the fundamental class of B. It is immediate that d_i is zero on $H^{n+1}(E;Z_2)$ (Chapter 7), and so $d_{i+1}p^*(v)$ is defined. Now consider the Serre exact sequence of this fibre space, with coefficients in Z_2 (the coefficients are suppressed for clarity):

$$\begin{array}{ccccccccccccc} 0 \to H^n(B) & \xrightarrow{p^*} & H^n(E) & \xrightarrow{j^*} & H^n(F) & \xrightarrow{\tau} & H^{n+1}(B) & \xrightarrow{p^*} & H^{n+1}(E) & \xrightarrow{j^*} & H^{n+1}F) \\ v & & \iota_E & & u & & d_i(v) & & d_{i+1}(\iota_E) & & d_1(u) \end{array}$$

Each group written here is isomorphic to Z_2 and is generated by the element written beneath it, according to the results of Chapter 9. By exactness, the first p^* is monomorphic and hence isomorphic, so that $p^*(v) = \iota_E$. Then the first j^* is zero, by exactness; so τ is isomorphic and $\tau(u) = d_i(v)$ in $H^{n+1}(B;Z_2)$. Thus the example meets the hypotheses of the theorem. Since τ is isomorphic, the second p^* is zero and the second j^* is isomorphic, so that

$$j^* d_{i+1} p^*(v) = j^* d_{i+1}(\iota_E) = d_1(u)$$

which is the assertion of the theorem for this case. In this special case there is zero indeterminancy, that is, the subgroup to be factored out is trivial, since $d_r = 0$ on $H^{n+1}(E;Z_2)$ for $r \leq i$.

Before proceeding with the proof of the general case, we alert the reader to the kind of language which will be used when dealing with maps into $K(\pi,n)$. Consider the following situation:

$$X \xrightarrow{f} Y \xrightarrow{g} K = K(\pi,n)$$

According to Theorem 1 of Chapter 1, there is a certain natural correspondence between homotopy classes of maps $Y \rightarrow K$ and elements of $H^n(Y;\pi)$. By virtue of the identifications

$$[g] \leftrightarrow g^*(\iota_n)$$
$$[g \circ f] \leftrightarrow (g \circ f)^*(\iota_n) = f^*(g^*(\iota_n))$$

and also because of the tendency to speak of g when we really should speak of the homotopy class $[g]$, we will be led to make such statements as

$$gf = f^*(g)$$

In the long run, this abuse of language will serve us well and we find it no more confusing than the cumbersome formulas which result if one meticulously preserves the distinctions between g, $[g]$, and $g^*(\iota_n)$.

Consider now the general case of the theorem. We will now use E, B, F, p, j, u, v, etc., to denote the objects and maps in the general case, and we distinguish the corresponding objects and maps in the universal example by a subscript zero. We work with the following diagram:

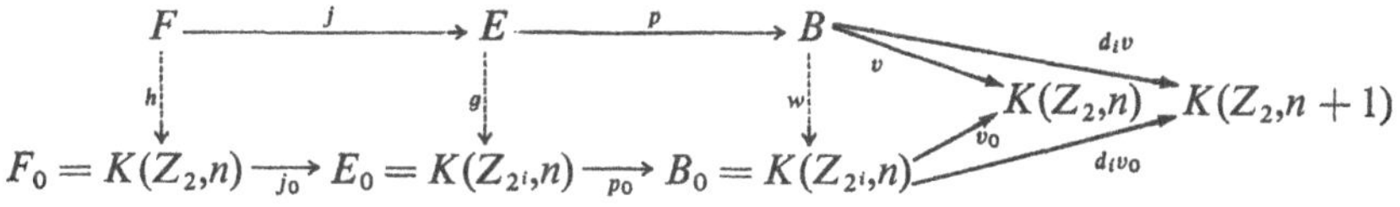

It is implicit in the hypotheses, since $d_i(v)$ is defined, that $d_j(v) = 0$ for all $j < i$, which means that $v \in H^n(B;Z_2)$ can be considered as the reduction mod 2 of a class $w \in H^n(B;Z_{2^i})$. This class is represented in the diagram as a map $B \rightarrow B_0$. The resulting triangle is commutative, $v_0 w \simeq v$, since v_0 is simply reduction mod 2. It is also clear that the other "triangle" is commutative: $(d_i v_0)w \simeq d_i v$; for these two maps obviously have the same effect on the fundamental class of $K(Z_2, n+1)$.

Now $(d_i v_0) \circ w \circ p \simeq (d_i v) \circ p = p^*(d_i v)$ and $d_i v = \tau(u)$, so that $p^*(d_i v) = p^*(\tau(u)) = 0$. But the universal example is induced by the map $d_i v_0$, and so $(d_i v_0) \circ w \circ p \simeq 0$ implies that the composite $w \circ p$ can be lifted to a map

$g: E \to E_0$ as shown in the diagram, and the square is strictly commutative: $wp = p_0 g$. Since pj is a constant map, the restriction of g to F, that is, gj, is a map into F_0, which we denote by h, making the left square strictly commutative also.

It would be nice if h were exactly u. We do not know this, but we have $\tau(h) = w^*(\tau_0(\iota_{F_0}))$ by naturality and we know $\tau_0(\iota_{F_0}) = d_i v_0$ from Formula 1 and Lemma 1; and therefore

$$\tau(h) = w^*(d_i v_0) = (d_i v_0)w \simeq d_i v = \tau(u)$$

so that at least $\tau(h) = \tau(u)$; this will be enough.

From the universal example, $j_0{}^* d_{i+1} p_0{}^* v_0 = d_1 u_0$. Applying h^* to this relation, using commutativity of the diagram and naturality of the Bockstein operators, we deduce that $j^* d_{i+1} p^* v = d_1 h$; but $d_1 h \equiv d_1 u$ modulo $d_1(\ker \tau) = d_1(\operatorname{im} j^*)$. This proves the theorem.

PRINCIPAL FIBRE SPACES

We now consider fibre spaces having certain additional structure. This sort of fibre space will be useful in Chapter 13.

Let $(E,p,B; F)$ be a fibre space (in the sense of Serre, as usual) and denote by E^* the space $\{(e_1,e_2): pe_1 = pe_2\}$, topologized as a subspace of $E \times E$. We say that $(E,p,B; F)$ is a *principal fibre space* if there are maps $\varphi: E \times F \to E$ and $h: E^* \to F$ such that:

1. φ acts fibrewise, i.e., the following diagram is commutative:

$$\begin{array}{ccc} E \times F & \xrightarrow{\varphi} & E \\ {\scriptstyle (p,\, id)}\downarrow & & \downarrow{\scriptstyle p} \\ B \times F & \xrightarrow[\pi_1]{} & B \end{array}$$

2. The restriction of φ to $F \times F$, where $F = p^{-1}(b_0)$ is a fixed fibre (b_0 the base point of B), is a map $F \times F \to F$ which, considered as a multiplication map, has a two-sided unit and a two-sided homotopy inverse
3. The composition

   $$E^* \xrightarrow{(\pi_1,h)} E \times F \xrightarrow{\varphi} E$$

   is homotopic to the projection π_2

For example, let E be the space of Moore paths over B, i.e., the space of paths (f,r) where r is a non-negative real number and f is a map of the closed interval $[0,r]$ into B, with $f(0) = b_0$. We have a natural projection

$p: E \to B$ given by $p(f,r) = f(r)$. Then the fibre $F = p^{-1}(b_0)$ is the space of Moore loops in B. (All spaces of functions are given the compact-open topology.) We obtain a map $\varphi: E \times F \to E$ by letting $\varphi(e,f)$ be the path obtained by first going around the loop f and then along the path e. We obtain a map $h: E^* \to F$ by letting $h(e_1,e_2)$ be the loop which goes out along e_2 and returns along e_1^{-1}. It is easy to verify that the three conditions are satisfied so that $(E,p,B; F)$ is a principal fibre space. For instance, condition (3) becomes the statement that $(e_2e_1^{-1})e_1$ is homotopic to e_2 (where e_1, e_2 are paths in B).

The reader will note that any principal fibre bundle in the sense of Steenrod is a principal fibre space in the sense above.

If $(E,p,B; F)$ is a principal fibre space, then the fibre space induced by a map of another space X into B is a principal fibre space. The proof is straightforward and is left to the reader (see Exercise 4).

The importance of principal fibre spaces lies in the following result.

Proposition 2

Let $(E,p,B; F)$ be a principal fibre space, and let v, v' be two maps of a space X into E. Then the composite maps pv, pv' are homotopic if and only if there exists a map $w: X \to F$ such that the composite $\varphi \circ (v,w)$ is homotopic to v'.

PROOF: Given w as above, since $p\varphi = \pi_1(p,id)$ by condition (1), we have

$$pv' \simeq p\varphi(v,w) \simeq \pi_1(p,id)(v,w) \simeq pv$$

Conversely, if it is given that $pv \simeq pv'$, we can use a covering homotopy to replace v' by another map in the same homotopy class such that $pv = pv'$. Then $(v,v'): X \to E \times E$ is a map of X into E^* and the composite $w = h(v,v'): X \to F$ has the required property.

DISCUSSION

The characteristic class of a fibre space has other interpretations. In particular, the reader is advised to find out the meaning (and proof) of the statement that "the characteristic class $\chi(\xi)$ measures the first obstruction to a cross section of ξ, assuming $\pi_1(B) = 0$."

The innocuous-looking Bockstein lemma (Theorem 1) will be crucial in the forthcoming calculations. We will prove a generalization of this lemma in Chapter 16 (Theorem 3). However, even in its ungeneralized form, this lemma played an important role in the history of calculation of homotopy groups; more accurately, failure to understand this lemma (and its mod p analogue) led for a time to some serious confusion in the field.

With this chapter we have at last completed our preparations; in Chapter 12 we compute.

EXERCISES

1. Verify that if ξ is a fibre space (in the sense of Serre or Hurewicz), then the "induced fibre space" $f^*(\xi)$ is actually a fibre space (in the same sense).
2. Show that $f^*(\xi)$ depends essentially only on the homotopy class of f, not on the choice of f within that class; in other words, by varying f within its homotopy class, we obtain fibre spaces which are all in the same fibre homotopy equivalence class.
3. Show that the characteristic class of a fibre space is a fibre homotopy invariant.
4. Verify that if (E,p,B) is a principal fibre space and if f is a map into B, then the fibre space induced by f is a principal fibre space.

REFERENCES

General properties of fibre spaces

1. E. Fadell [1].
2. P. J. Hilton [1].
3. S.-T. Hu [1].

Cross-sections of fibre bundles

1. N. E. Steenrod [1].

Principal fibre spaces

1. J.-P. Meyer [1].
2. F. P. Peterson and N. Stein [1].

CHAPTER 12

APPLICATIONS: SOME HOMOTOPY GROUPS OF SPHERES

Using the machinery which has been developed in the last few chapters, we can now make some computations in the homotopy groups of spheres.

The idea behind the technique is quite simple. To obtain the 2-components of the homotopy groups of S^n, we begin with $K(Z,n)$, which has the same cohomology and homotopy groups as S^n up through dimension n. However, $K(Z,n)$ has non-zero cohomology in higher dimensions, whereas S^n has non-zero cohomology only in dimension n. We will modify $K(Z,n)$ so as to eliminate or "kill" the non-zero mod 2 cohomology classes in dimensions $n+1, n+2, \ldots$, and then the 2-components of the homotopy groups of the resulting space will be the same as those of the sphere in those dimensions, by the $\mathcal{C}_p$ approximation theorem (Theorem 4 of Chapter 10).

THE SUSPENSION THEOREM

It makes sense to talk about "the homotopy groups of S^n" because to a great extent these groups are independent of n. To be precise, $\pi_{n+k}(S^n)$ is independent of n provided only that $n \geq k+2$. This follows from the Freudenthal suspension theorem. We will indicate the proof of this standard result in homotopy. For convenience we suppose that each space X comes provided with a distinguished "base point," denoted by an asterisk (*). The *reduced suspension* of X, SX, is obtained from $X \times I$ (where I as usual is the unit interval) by collapsing to a single point the subspace

$$(X \times 0) \cup (X \times 1) \cup (* \times I)$$

Then it is well known that $\tilde{H}_i(X) \approx \tilde{H}_{i+1}(SX)$ (where $\tilde{H}$ denotes the reduced homology). There is a natural map $\varphi\colon X \to \Omega SX$, given by defining $\varphi(x)$ as the path $t \to (x,t)$, in a natural notation. Then the *suspension homomorphism* E (for Einhängung) is defined by the diagram below. In fact E can be defined more generally as a homomorphism of $[Y,X]$ into $[SY,SX]$ as a homomorphism, that is, whenever the set of homotopy classes $[Y,X]$ has a natural group structure).

$$\begin{array}{ccc} \pi_i(X) & \xrightarrow{\varphi_*} & \pi_i(\Omega SX) \\ & \searrow^{E} & \downarrow \approx \\ & & \pi_{i+1}(SX) \end{array}$$

Now suppose that X is $(n-1)$-connected; then it follows that SX is n-connected and ΩSX is $(n-1)$-connected. If LX is the standard contractible fibre space of paths in X, then the Serre homology exact sequence of this fibre space begins as follows:

$$\begin{array}{ccccccc} H_{2n}(LX) & \to & H_{2n}(SX) & \xrightarrow{\tau} & H_{2n-1}(\Omega SX) & \to & H_{2n-1}(LX) \\ & & \uparrow \approx & \nearrow_{\varphi} & & & \\ & & H_{2n-1}(X) & & & & \end{array}$$

where the groups on the ends are zero since LX is contractible. It can be shown that τ and φ are compatible, i.e., that the triangle in the diagram is commutative, and thus φ induces isomorphisms on H_k for all $k < 2n-1$ and an epimorphism for H_{2n-1}. But then the corresponding assertions hold for the induced maps in homotopy, namely, for the suspension homomorphism E. Thus we have indicated the proof of the following important result.

Theorem 1 (suspension theorem)

If X is $(n-1)$-connected, then the suspension homomorphism $E\colon \pi_i(X) \to \pi_{i+1}(SX)$ is isomorphic for $i < 2n-1$ and epimorphic for $i \leq 2n-1$.

In particular, taking $X = S^n$ and observing that $S(S^k) = S^{k+1}$, we have the previously stated corollary that $\pi_{n+k}(S^n)$ is independent of n for $n \geq k+2$.

A BETTER APPROXIMATION TO S^n

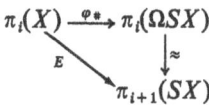

Throughout the remainder of this chapter, when we say "homotopy group," it is understood that we mean the 2-component of the homotopy group; the symbol $\pi_n(X)$ is to be interpreted in the same way. Cohomology will be with Z_2 coefficients.

Recall that $H^{n+1}(Z,n; Z_2) = 0$ and that $H^{n+2}(Z,n; Z_2) = Z_2$ generated by $Sq^2(\iota_n)$ where ι_n is the fundamental class. This cohomology class defines a map (or rather a homotopy class)

$$Sq^2: K(Z,n) \to K(Z_2, n+2)$$

by Theorem 1 of Chapter 1. From the standard contractible fibre space over $K(Z_2, n+2)$, this map induces a fibre space X_1 over $K(Z,n)$ with fibre $\Omega K(Z_2, n+2) = K(Z_2, n+1)$. We view X_1 as a better approximation to S^n than $K(Z,n)$. We say this because $H^{n+2}(X_1;Z_2)$ is zero, the class $Sq^2 \in H^{n+2}(Z,n; Z_2)$ having been killed. This is verified by considering the cohomology exact sequence of the fibre space $F_1 \to X_1 \to B$, where $F_1 = K(Z_2, n+1)$ and $B = K(Z,n)$. By construction, the fundamental class ι_{n+1} of F_1 transgresses to $Sq^2(\iota_n)$. Then $Sq^1(\iota_{n+1})$, which generates $H^{n+2}(F_1)$, transgresses to $Sq^1Sq^2(\iota_n) = Sq^3(\iota_n)$. Thus, in the Serre cohomology sequence (coefficients in Z_2),

$$\cdots H^{n+1}(F_1) \xrightarrow{\tau} H^{n+2}(B) \xrightarrow{p^*} H^{n+2}(X_1) \xrightarrow{i^*} H^{n+2}(F_1) \xrightarrow{\tau} H^{n+3}(B) \cdots$$

the transgression τ is onto $H^{n+2}(B)$ and monomorphic on $H^{n+2}(F_1)$, so that, by exactness, p^* and i^* are both zero in the dimension shown and thus $H^{n+2}(X_1) = 0$. Note that the method requires not only that the fundamental class of the fibre transgress to the appropriate class in the base but also that the transgression be monomorphic on the next dimension of the fibre.

Now let $f: S^n \to K(Z,n)$ represent the homotopy class of a generator of $\pi_n(K(Z,n)) = Z$. Then, in cohomology, f^* is isomorphic through dimension n and monomorphic in dimension $n+1$—the last is obvious because $H^{n+1}(Z,n; Z_2) = 0$. Therefore, in homotopy, $f_\#$ is isomorphic through π_n and epimorphic on π_{n+1} (mod $\mathcal{C}_2$), by the $\mathcal{C}_p$ approximation theorem. This much is not new. But the composition

$$S^n \xrightarrow{f} K(Z,n) \xrightarrow{Sq^2} K(Z_2, n+2)$$

is null-homotopic, since $\pi_n(K(Z_2, n+2)) = 0$. Therefore f may be lifted to a map $f_1: S^n \to X_1$. Now f_1 meets the hypotheses of the theorem with $n+1$ in place of n, and we can conclude that f_1 induces a $\mathcal{C}_2$-isomorphism on π_{n+1}. But the homotopy groups of X_1 are easily obtained from the homotopy exact sequence of the fibre space: $\pi_n(X_1) = Z$, $\pi_{n+1}(X_1) = Z_2$, and all other $\pi_m(X_1)$ are zero. In particular, $\pi_{n+1}(S^n)$ must be Z_2.

The next step is to kill $H^{n+3}(X_1)$, obtaining a fibre space $F_2 \to X_2 \to X_1$ where the total space X_2 will have zero cohomology in dimension $n+3$ and therefore will have the same homotopy group as S^n in dimension $n+2$.

CALCULATION OF $\pi_{n+k}(S^n)$, $k \leq 7$

We will compute $\pi_{n+k}(S^n)$ for $k \leq 7$, assuming n large. Precisely, since we will obtain $H^{n+k}(X_1)$ up to $k = 11$, we will want to assume n comfortably larger than 11, in order that the cohomology of the $K(\pi,n)$ spaces which occur will not contain any cup products in the range under examination.

We have already begun. The first step is the construction of the induced fibre space X_1, where in the diagram, L denotes the contractible space of

$$\begin{array}{ccc} F_1 = K(Z_2, n+1) \xrightarrow{i} X_1 & & L \\ \quad\quad\quad\quad\quad\quad\quad \downarrow p & & \downarrow \\ \quad\quad\quad B = K(Z,n) & \xrightarrow{Sq^2} & K(Z_2, n+2) \end{array}$$

paths in $K(Z_2, n+2)$. We can compute the cohomology of X_1 using the long exact sequence of Serre, as follows.

$$\begin{array}{ccc} H^*(Z_2, n+1) & \xleftarrow{i^*} & H^*(X_1) \\ & \searrow^{\tau} & \uparrow p^* \\ & & H^*(Z,n) \end{array}$$

We have already obtained $H^{n+2}(X_1) = 0$ in this manner, but we will need $H^{n+k}(X_1)$ with $k \leq 11$. Observe that elements of $H^*(X_1)$ may be regarded as being of two kinds: if $i^*(x) = 0$, then $x \in \operatorname{im}(p^*)$, or in fact $x \in p^*(\operatorname{coker} \tau)$; otherwise, $i^*(x) \neq 0$ so that $x \in (i^*)^{-1}(\ker \tau)$. Thus in order to compute $H^*(X_1)$, it is sufficient to know $H^*(F_1)$, $H^*(B)$, and the transgression.

This presents no difficulties, since $H^*(F_1)$ and $H^*(B)$ are known (Chapter 9), and the transgression is given by $\tau(\iota_{n+1}) = Sq^2(\iota_n)$ and the fact that τ commutes with the squaring operations. Thus we have

$$\tau(Sq^1\iota_{n+1}) = Sq^1\tau(\iota_{n+1}) = Sq^1 Sq^2(\iota_n) = Sq^3(\iota_n)$$
$$\tau(Sq^2\iota_{n+1}) = Sq^2\tau(\iota_{n+1}) = Sq^2 Sq^2(\iota_n) = Sq^3 Sq^1(\iota_n) = Sq^3(0) = 0$$

and so forth.

The calculation of $\tau\colon H^j(F_1) \to H^{j+1}(B)$ is presented, in a condensed notation, in the first two columns of the table on page 116. This notation is to be read as shown in the small table here. The cokernel of τ contains

Condensed statement	Full statement
$2 \leftarrow \iota_{n+1}$	$\tau(\iota_{n+1}) = Sq^2(\iota_n)$
$3 \leftarrow 1$	$\tau(Sq^1(\iota_{n+1})) = Sq^3(\iota_n)$
$\left.\begin{matrix} 9 \\ 7,2 \end{matrix}\right\} \leftarrow 4,2,1$	$\tau(Sq^4 Sq^2 Sq^1(\iota_{n+1})) = Sq^9(\iota_n) + Sq^7 Sq^2(\iota_n)$

elements such as $Sq^4(\iota_n)$, represented by a "4" in the first column of the table on page 116, not at the end of any arrow; the kernel of τ contains elements such as $Sq^2(\iota_{n+1})$, represented by the "2" in the second column and also by such elements as $Sq^5(\iota_{n+1}) + Sq^4Sq^1(\iota_{n+1})$, where each summand has the same image under τ, shown in the table by the common destination of the arrows originating at the "5" and at the "4,1" in the second column.

The reader should be able to reproduce this calculation of τ without peeking. To write down $H^*(Z,n; Z_2)$ and $H^*(Z_2, n+1; Z_2)$ requires only a knowledge of the basis for the Steenrod algebra given in Chapter 3, together with the results of Chapter 9. The calculation of the transgression will give useful practice in the Adem relations.

Note that this part of the calculation could be continued indefinitely, if we assume n sufficiently large. The description of $H^*(F_1)$ and $H^*(B)$, and the exact sequence itself, are valid up to the neighborhood of dimension $2n$.

From this calculation of τ, we can write down a basis for the cohomology of X_1. $H^n(X_1)$ is generated by a fundamental class ι_n which is the image of the fundamental class of B under p^*; we use the same notation for both these classes. $H^{n+3}(X_1)$ is Z_2 generated by a class α such that $i^*(\alpha) = Sq^2(\iota_{n+1})$. The general situation is represented by $H^{n+4}(X_1)$. Here we have a class $p^*(Sq^4(\iota_n))$ (we will drop the p^* from our notation), and we also must have a class β such that $i^*(\beta) = Sq^3(\iota_{n+1})$. Notice that there is some indeterminacy in the choice of β, since $i^*(p^*(Sq^4(\iota_n))) = 0$. One might try to identify a canonical choice of β, for instance by relating it to $Sq^1(\alpha)$. We will not do this; we will simply assume that β is some class, arbitrarily chosen subject to the condition $i^*(\beta) = Sq^3(\iota_n)$; whatever statements we make about β are independent of the choice.

In the third column of the table on page 116, a basis for $H^{n+k}(X_1)$ has been written for $k \leq 11$. In dimension $n + 10$, for example, there is a class $p^*(Sq^{10}(\iota_n))$, which we write simply as $Sq^{10}(\iota_n)$ (consistent with our writing ι_n for $p^*(\iota_n)$). Then there are three other generators, each of which has a prescribed image under i^*; each of these elements (denoted λ, μ, ν in the table) could be chosen in two different ways, since $i^*(Sq^{10}(\iota_n)) = 0$.

As we continue, we will have need of certain Bockstein relations in $H^*(X_1)$, which we set forth as a lemma.

Lemma 1

In $H^*(X_1)$ (coefficients Z_2 as always),

1. $d_1(\alpha) = \beta + (\ )Sq^4\iota_n$ where ( ) denotes an undetermined coefficient
2. $d_2(Sq^4\iota_n) = \gamma$
3. $d_1(\gamma) = 0$
4. $d_2(Sq^8\iota_n) = \kappa$

k	$H^{n+k}(Z,n;Z_2)$	$H^{n+k}(Z_2,n+1;Z_2)$	$H^{n+k}(X_1;Z_2)$	$H^{n+k}(Z_2,n+2;Z_2)$
0	ι_n		ι_n	
1		ι_{n+1}		
2	2	1		ι_{n+2}
3	3	2	$\alpha(2)$	1
4	4	3	$Sq^4\iota_n$	2
		2,1	$\beta(3)$	
5	5	3,1	$\gamma(3,1)$	3
		4		2,1
6	6	5	$Sq^6\iota_n$	3,1
	4,2	4,1	$\delta(5+4,1)$	4
7	7	6	$Sq^7\iota_n$	5
	5,2	5,1	$\varepsilon(5,1)$	4,1
		4,2	$\zeta(4,2)$	
8	8	4,2,1	$Sq^8\iota_n$	6
	6,2	7	$\eta(5,2)$	5,1
		6,1		4,2
		5,2		
9	9	8	$\theta(6,2)$	7
	7,2	7,1	$\kappa(5,2,1)$	4,2,1
	6,3	6,2		6,1
		5,2,1		5,2
10	10	9	$Sq^{10}\iota_n$	8
	8,2	6,2,1	$\lambda(7,2)$	7,1
	7,3	8,1	$\mu(9+8,1+6,2,1)$	6,2
		7,2	$\nu(6,3)$	5,2,1
		6,3		
11	11	10	$Sq^{11}\iota_n$	
	9,2	9,1	$\xi(8,2)$	
	8,3	8,2	$\pi(7,3)$	
		7,3	$\rho(6,3,1)$	
		7,2,1	$\sigma(9,1+7,2,1)$	
		6,3,1		
12	12			
	10,2			
	9,3			
	8,4			

k	$H^{n+k}(X_2;Z_2)$	$H^{n+k}(Z_8, n+3; Z_2)$	$H^{n+k}(X_3;Z_2)$	$H^{n+k}(Z_2, n+6)$
0	ι_n		ι_n	
1				
2				
3		ι_{n+3}		
4	$Sq^4\iota_n$	$d_3\iota_{n+3}$		
5	$A(3)$	2		
6	$Sq^6\iota_n$	3		ι_{n+6}
		$2d_3$		
7	$Sq^7\iota_n$	4	$P(4)$	1
	$B(5+4,1)$	$3d_3$		
8	$Sq^8\iota_n$	5	$Sq^8\iota_n$	2
	$C(5,1)$	$4d_3$	$Q(5)$	
9	$D(5,2)$	6	$R(6)$	?
		$5d_3$	$S(5d_3)$	
		4,2		
10	$Sq^{10}\iota_n$			

k	$H^{n+k}(X_4;Z_2)$	$H^{n+k}(Z_{16}, n+7)$	$H^{n+k}(X_5;Z_2)$
0	ι_n		ι_n
...	0		0
7		ι_{n+7}	0
8	$Sq^8\iota_n$	d_4	0
9	S'		?
	$+?$		

The notation is that of the table.

PROOF:

1. We have $i^*(\alpha) = Sq^2\iota_{n+1}$ by definition of α. Therefore $i^*(d_1\alpha) = d_1 i^*(\alpha) = d_1 Sq^2\iota_{n+1} = Sq^3\iota_{n+1}$. We have also $i^*(\beta) = Sq^3\iota_{n+1}$. Since $Sq^4\iota_n$ is in the image of p^*, $i^*Sq^4\iota_n = 0$. This proves item (1).
2. We have $d_1Sq^4\iota_n = Sq^1Sq^4\iota_n = Sq^5\iota_n = \tau(Sq^2Sq^1\iota_{n+1})$. Thus $d_1p^*Sq^4\iota_n = 0$ since $p^*\tau = 0$, and, by the "Bockstein lemma" (Theorem 1 of Chapter 11), $i^*d_2p^*Sq^4\iota_n = d_1(Sq^2Sq^1\iota_{n+1}) = Sq^3Sq^1\iota_{n+1} = i^*(\gamma)$. Since γ is the only generator in $H^{n+5}(E)$, the result follows.
3. This follows from (2) since $d_1 = 0$ on anything in the image of any d_r.
4. Again, as in (2), we use the Bockstein lemma. Since $d_1Sq^8\iota_n = Sq^9\iota_n = \tau(Sq^7 + Sq^4Sq^2Sq^1)(\iota_{n+1})$, we have

$$i^*d_2p^*Sq^8\iota_n = d_1(Sq^7 + Sq^4Sq^2Sq^1)(\iota_{n+1}) = Sq^5Sq^2Sq^1(\iota_{n+1}) = i^*(\kappa)$$

Since $i^*(\theta) = Sq^6 Sq^2 \iota_{n+1}$ and there are no other generators in this dimension, the result follows.

THE CALCULATION CONTINUES

We have constructed the space X_1 so that $H^*(X_1)$ and $H^*(S^n)$ (with Z_2 coefficients) agree through dimension $n+2$, and we have a map $f_1: S^n \to X_1$ inducing the isomorphisms. The next step is to kill the class $\alpha \in H^{n+3}(X_1)$ in such a way as to produce a space X_2 with no cohomology in dimension $n+3$. Recall that this requires killing α in such a way that the transgression is non-zero on the first class in the fibre above the fundamental class. This can be accomplished by taking X_2 as the induced fibre space over X_1 by a map representing α where the new fibre is $K(Z_2, n+2)$, so that the funda-

$$\begin{array}{ccc} F_2 = K(Z_2, n+2) \longrightarrow & X_2 & \\ & \downarrow & \\ & X_1 & \xrightarrow{\alpha} K(Z_2, n+3) \end{array}$$

mental class transgresses to α and Sq^1 of the fundamental class transgresses to $Sq^1(\alpha)$, which is non-zero by part (1) of Lemma 1. If indeed $H^{n+3}(X_2)=0$, then X_2 is an improvement over X_1 as an approximation to the sphere; we have $\pi_{n+2}(X_2) = Z_2$ from the homotopy exact sequence of the fibre space $F_2 \to X_2 \to X_1$, and therefore $\pi_{n+2}(S^n) = Z_2$.

In order to obtain the cohomology of X_2, we must compute the transgression in the cohomology of the fibre space $F_2 \to X_2 \to X_1$. This will be less routine than the analogous step in the previous stage, because now we do not know precisely the squaring operations in $H^*(X_1)$. However, we know enough to obtain a basis for $H^*(X_2)$ for a considerable range.

The calculation of this transgression is indicated roughly in the third and fourth columns of the table on page 116. By construction, the fundamental class ι_{n+2} transgresses to α. Then, as remarked already, $Sq^1(\iota_{n+2})$ transgresses to $Sq^1(\alpha) = d_1(\alpha)$, which by Lemma 1 is β plus possibly $Sq^4(\iota_n)$. It is not essential to know whether or not this second term is present in $Sq^1(\alpha)$ (this depends on the choice of β) because either way we obtain a single generator for $H^{n+4}(X_2)$ from the cokernel of τ, and this class is $p_2^*(Sq^4\iota_n)$ where $p_2: X_2 \to X_1$. Therefore the first class in $H^*(X_2)$, after the fundamental class, will be this one, which we denote by $Sq^4\iota_n$, dropping the p_2^*. In fact this class may equally well be regarded as Sq^4 on the fundamental class of X_2, using the naturality of Sq^4 with respect to p_2.

Now $Sq^2(\iota_{n+2})$ transgresses to $Sq^2(\alpha)$. We must determine whether $Sq^2(\alpha)$ is γ or zero. But $i^*(\alpha)=Sq^2(\iota_{n+1})$, by definition of α; then $i^*Sq^2\alpha=Sq^2i^*\alpha=Sq^2Sq^2\iota_{n+1}=Sq^3Sq^1\iota_{n+1}=i^*\gamma$. This proves that $Sq^2\alpha=\gamma$ or, what is essential for us, that $\tau(Sq^2\iota_{n+1})=\gamma$.

Similarly, $\tau(Sq^2Sq^1\iota_{n+2})=\delta$ plus possibly $Sq^6\iota_n$, since

$$\begin{aligned} i^*(Sq^2Sq^1\alpha)=Sq^2Sq^1(i^*\alpha)&=Sq^2Sq^1Sq^2\iota_{n+1}\\ &=(Sq^5+Sq^4Sq^1)\iota_{n+1}\\ &=i^*(\delta)\end{aligned}$$

so that $Sq^2Sq^1\alpha=\delta+(\ )Sq^6\iota_n$ where $(\ )$ is an undetermined coefficient. On the other hand, the same argument gives $\tau(Sq^3\iota_{n+2})=0+(\ )Sq^6\iota_n$. In order to know whether or not $\tau\colon H^{n+3}(F_2)\to H^{n+4}(X_1)$ is epimorphic, that is, in order to obtain its cokernel, we must get some information about these coefficients. In the present case we can argue as follows. We have $Sq^3\iota_{n+2}=Sq^1Sq^2\iota_{n+2}$, and therefore

$$\tau(Sq^3\iota_{n+2})=Sq^1(\tau(Sq^2\iota_{n+2}))=Sq^1\gamma$$

But $Sq^1(\gamma)=0$ by part (3) of Lemma 1. Thus $\tau(Sq^3\iota_{n+2})=0$. This is enough, since it now follows, regardless of the coefficient in $\tau(Sq^2Sq^1\iota_{n+2})$, that the cokernel of τ in this dimension is generated by $Sq^6\iota_n$ (dropping the p_2^* as usual).

In the next dimension, similar arguments give

$$\begin{aligned}\tau(Sq^4\iota_{n+2})&=\zeta+(\ )Sq^7\iota_n\\ \tau(Sq^3Sq^1\iota_{n+2})&=\varepsilon+(\ )Sq^7\iota_n\end{aligned}$$

so that the cokernel is generated by $Sq^7\iota_n$ (regardless of the unknown coefficients).

In the next dimension we have

$$\begin{aligned}\tau(Sq^5\iota_{n+2})&=\eta+(\ )Sq^8\iota_n\\ \tau(Sq^4Sq^1\iota_{n+2})&=\eta+(\ )Sq^8\iota_n\end{aligned}$$

and we must determine whether the coefficients are the same. But $Sq^5+Sq^4Sq^1=Sq^2Sq^3$, by the Adem relation; and we have seen that $\tau(Sq^3\iota_{n+2})=0$, from which it follows that

$$\begin{aligned}\tau(Sq^5\iota_{n+2})+\tau(Sq^4Sq^1\iota_{n+2})&=\tau(Sq^2Sq^3\iota_{n+2})\\ &=Sq^2(\tau(Sq^3\iota_{n+2}))\\ &=0\end{aligned}$$

which is all we need to know. It shows that the kernel of τ contains $(Sq^5+Sq^4Sq^1)(\iota_{n+2})$ and that the cokernel contains $Sq^8\iota_n$.

By similar reasoning, we can complete a basis for $H^*(X_2)$ at least through dimension $n+10$; this is given in the first column on the top of page 117. Perhaps the only other subtlety is showing that $\tau(Sq^5 Sq^2 \iota_{n+2}) = 0$. This follows from $Sq^5 Sq^2 = Sq^1(Sq^4 Sq^2)$ and from part (4) of Lemma 1, since $d_1 d_2 = 0$.

This calculation has verified that $H^{n+3}(X_2) = 0$ as desired, and thus $\pi_{n+2}(S^n) = Z_2$. The next step will be to kill the class $Sq^4 \iota_n \in H^{n+4}(X_2)$. We will need to know the following Bockstein relations.

Lemma 2

In $H^*(X_2)$, (1) $d_3 Sq^4 \iota_n = A$, and (2) $d_3 Sq^8 \iota_n = D$. Here A and D are defined by $i^* A = Sq^3 \iota_{n+2}$ and $i^* D = Sq^5 Sq^2 \iota_{n+2}$ (as indicated in the table).

PROOF: In both cases the proof is based on the Bockstein lemma of Chapter 11. In (1) we have, by part (2) of Lemma 1,

$$d_2 Sq^4 \iota_n = \gamma = \tau(Sq^2 \iota_{n+2}) \in H^{n+5}(X_1)$$

and therefore

$$i^*(d_3 Sq^4 \iota_n) = d_1(Sq^2 \iota_{n+2}) = Sq^3 \iota_{n+2} = i^* A$$

which implies the result. In the same way, part (2) follows from part (4) of Lemma 1.

Now our objective is to kill $H^{n+4}(X_2)$. This means killing the class $Sq^4 \iota_n$. This class can be realized by a map of X_2 into $K(Z_2, n+4)$, but using this map to induce a fibre space over X_2 will not do the trick, because the fibre will have a class $Sq^1 \iota_{n+3}$ which will transgress to *zero* since $d_1 Sq^4 \iota_n = 0$. The fact that d_1 and d_2 are zero on $Sq^4 \iota_n$ implies that this class is the reduction (mod 2) of a class with Z_8 coefficients. The proper procedure is to map X_2 into $K(Z_8, n+4)$ by a map corresponding to a class which reduces to $Sq^4 \iota_n$ (mod 2). Then H^{n+4} of the new fibre is generated by d_3 of the fundamental class, which will transgress to $d_3 Sq^4 \iota_n = A$.

Thus we construct a fibre space induced from the standard contractible

$$\begin{array}{ccc} F_3 = K(Z_8, n+3) \longrightarrow & X_3 & \\ & \downarrow & \\ & X_2 \longrightarrow & K(Z_8, n+4) \end{array}$$

fibre space over $K(Z_8, n+4)$ by a map having the properties described, which we might, by abuse of language, refer to as "$Sq^4 \iota_n$" for heuristic reasons.

To obtain $H^*(X_3)$, we must calculate the transgression in this fibre space. This calculation is indicated in the first two columns, top of page 117. Most of the details are based on types of arguments which have been used before. We mention that $\tau(Sq^4\iota_{n+3})=0$ because $Sq^4Sq^4\iota_n = Sq^7Sq^1\iota_n + Sq^6Sq^2\iota_n$, and Sq^1 and Sq^2 are obviously zero on the fundamental class ι_n of X_2 since $H^*(X_2)=0$ in the relevant dimensions. As a corollary, $\tau(Sq^5\iota_{n+3})$ is also zero. In a similar fashion $\tau(Sq^6\iota_{n+3})=0$ by the Adem relation $Sq^6Sq^4 = Sq^7Sq^3$. The fact that $\tau(Sq^5d_3\iota_{n+3})=0$ follows from the fact that $Sq^1D = d_1d_3Sq^8\iota_n = 0$. The other details are left to the reader.

Now we can write a basis for $H^{n+k}(X_3)$ for $k \leq 9$, as done in column 3, top of page 117. Notice that we have killed not only $H^{n+4}(X_2)$ but also H^{n+5} and H^{n+6}. This reflects the fact that $\pi_{n+4}(S^n)$ and $\pi_{n+5}(S^n)$ are zero.

We must kill the class $P \in H^{n+7}(X_3)$, where $i^*P = Sq^4\iota_{n+3}$. Note that Sq^1P is non-zero, since $Sq^1Sq^4\iota_{n+3}$ is also in the kernel of the transgression; in the notation of the table, $Sq^1P = Q + (\ )Sq^8\iota_n$. Therefore we construct a fibre space

$$\begin{array}{ccc} F_4 = K(Z_2, n+6) \longrightarrow & X_4 & \\ & \downarrow & \\ & X_3 & \xrightarrow{P} K(Z_2, n+7) \end{array}$$

and this will kill off H^{n+7}. It is easy to obtain $H^{n+k}(X_4)$ for $k \leq 8$; in fact the only non-zero classes are ι_n and $Sq^8\iota_n$ in this range.

Lemma 3

In $H^{n+9}(X_4)$, $d_4(Sq^8\iota_n)$ is non-zero.

PROOF: We have shown (Lemma 2) that

$$d_3Sq^8\iota_n = D = \tau(Sq^4d_3\iota_{n+3}) \in H^{n+9}(X_2)$$

and therefore the Bockstein lemma implies that

$$\begin{aligned} i^*(d_4Sq^8\iota_n) &= d_1(Sq^4d_3\iota_{n+3}) \\ &= Sq^5d_3\iota_{n+3} \\ &= i^*S \end{aligned}$$

where S is an appropriate class in $H^{n+9}(X_3)$. Since S is clearly in the cokernel of $\tau\colon H^{n+8}(F_4) \to H^{n+9}(X_3)$, it gives rise to a class $S' \in H^{n+9}(X_4)$ and the result follows from the naturality of d_4.

It follows that we can kill $H^{n+8}(X_4)$ by using a map

$$X_4 \xrightarrow{\text{“}Sq^8\iota_n\text{”}} K(Z_{16}, n+8)$$

corresponding to a cohomology class in $H^{n+8}(X_4;Z_{16})$ which reduces to $Sq^8\iota_n$ (mod 2). This gives us a space X_5 which has the same cohomology as S^n through dimension $n+8$. Moreover it is easy to lift our map $f_1\colon S^n \to X_1$ up through the X_i to a map $f_5\colon S^n \to X^5$ which induces isomorphisms on cohomology through dimension $n+8$ and therefore induces isomorphisms in homotopy groups through π_{n+7}. The homotopy groups of X_5 are easily obtained using the homotopy exact sequences of the successive fibre spaces in the construction. These are the same as the homotopy groups of S^n through π_{n+7}. We can therefore state the following results:

Theorem 2

k	0	1	2	3	4	5	6	7
$\pi_{n+k}(S^n)$	Z	Z_2	Z_2	Z_8	0	0	Z_2	Z_{16}

Recall that what is tabulated here is the 2-component of the homotopy group. Also note that these results are valid for all n such that $n \geq k+2$; we have assumed n large in the calculation, but $\pi_{n+k}(S^n)$ is stable as soon as $n \geq k+2$, by the suspension theorem.

The cohomological techniques which we have used to compute homotopy groups cannot be extended indefinitely. Eventually one reaches an ambiguity which cannot be resolved by these methods. In particular, $\pi_{n+14}(S^n)$ must be settled in another way. In Chapter 18 we will develop more sophisticated algebraic techniques; these will help to explain the difficulties and the need for other methods.

Figure 1 displays the tower of fibre spaces which we have used in our calculations.

$$\begin{array}{lcl}
F_5 = K(Z_{16}, n+7) \longrightarrow & X_5 & \\
 & \downarrow & \\
F_4 = K(Z_2, n+6) \longrightarrow & X_4 & \xrightarrow{\text{“}Sq^8\iota_n\text{”}} K(Z_{16}, n+8) \\
 & \downarrow & \\
F_3 = K(Z_8, n+3) \longrightarrow & X_3 & \xrightarrow{P} K(Z_2, n+7) \\
 & \downarrow & \\
F_2 = K(Z_2, n+2) \longrightarrow & X_2 & \xrightarrow{\text{“}Sq^4\iota_n\text{”}} K(Z_8, n+4) \\
 & \downarrow & \\
F_1 = K(Z_2, n+1) \longrightarrow & X_1 & \xrightarrow{\alpha} K(Z_2, n+3) \\
 & \downarrow & \\
B = & K(Z,n) & \xrightarrow{Sq^2} K(Z_2, n+2)
\end{array}$$

Figure 1

DISCUSSION

Although the suspension theorem (Theorem 1) leads naturally to the question of $\pi_{n+k}(S^n)$ for n large (the kth *stable* homotopy group of spheres), progress at first was slow. Of course, the result for $k=0$ is immediate by the Hurewicz theorem; on the other hand, the value for $k = 2$ was in dispute for some time until settled by G. W. Whitehead (using geometric techniques).

Serre's results drastically changed the situation and facilitated many cohomological computations, including the ones given here. In Chapter 18 we shall refine the cohomological method and indicate how the results of Theorem 2 can easily be greatly extended.

The cohomological method is by no means the only possible approach to the problem. The so-called "composition method" in particular has been extensively used by Toda and Barratt. In Chapter 16 we shall interpret the results of Theorem 2 in terms of this method. However, we do not touch on two of their basic tools, the "EHP-sequence" and J-homomorphism of G. W. Whitehead.

We have sketched a homological proof of Theorem 1. For $X = S^n$, the theorem, due to Freudenthal, greatly antedates the proof we have given. A modern geometric proof in this case has been given by Bott.

APPENDIX: SOME HOMOTOPY GROUPS OF S^3

The computations just carried out for $\pi_{n+k}(S^n)$ depend on the assumption that n is large. We here present a sketch of the computation of $\pi_{3+k}(S^3)$ $(k \leq 3)$, in order to indicate what modifications and complications occur in the technique. As before, we deal with the 2-components and all cohomology is understood to be with Z_2 coefficients.

We take $B = K(Z,3)$ as our first approximation to S^3. We can write down $H^*(K(Z,3))$, using the results of Chapter 9; it is the polynomial algebra in the fundamental class ι_3 and in sequences $Sq^I(\iota_3)$ where I is admissible, has excess less than 3, and does not terminate in 1. Thus we do not find $Sq^4\iota_3$ or $Sq^5\iota_3$, etc., because the excess is too large; we find cup products such as $(\iota_3)(Sq^2\iota_3)$; and $Sq^3\iota_3$ appears as $(\iota_3)^2$.

Comparing $H^*(K(Z,3))$ with $H^*(S^3)$, we see that the first class to be killed is $Sq^2\iota_3$ (just as the first class to be killed in the stable case is $Sq^2\iota_n$). We therefore use a map

$$B = K(Z,3) \xrightarrow{Sq^2\iota_3} K(Z_2,5)$$

to construct a fibre space over B with total space X_1 and fibre $F_1 = K(Z_2,4)$. Notice that $d_1 Sq^2\iota_3 = Sq^3\iota_3 = (\iota_3)^2$, which is non-zero, so that Z_2 is the correct choice for the first homotopy group.

There is no difficulty in writing $H^*(F_1)$; it is a polynomial algebra in ι_4 and in $Sq^I\iota_4$ where I is admissible and has excess less than 4.

We next must calculate $H^*(X_1)$. This is done by means of Serre's cohomology spectral sequence. As stated in Chapter 8, the E_2 term of this spectral sequence is the tensor product $H^*(B)\otimes H^*(F_1)$. When n is large and k small, $H^{n+k}(K(Z,n))$ and $H^{n+k}(K(Z_2, n+1))$ do not contain any cup-product terms and the spectral sequence reduces to Serre's exact sequence; every element in this range of $H^*(F_1)$ is transgressive. But in the present case we must handle the spectral sequence in its general form. Figure 2 is

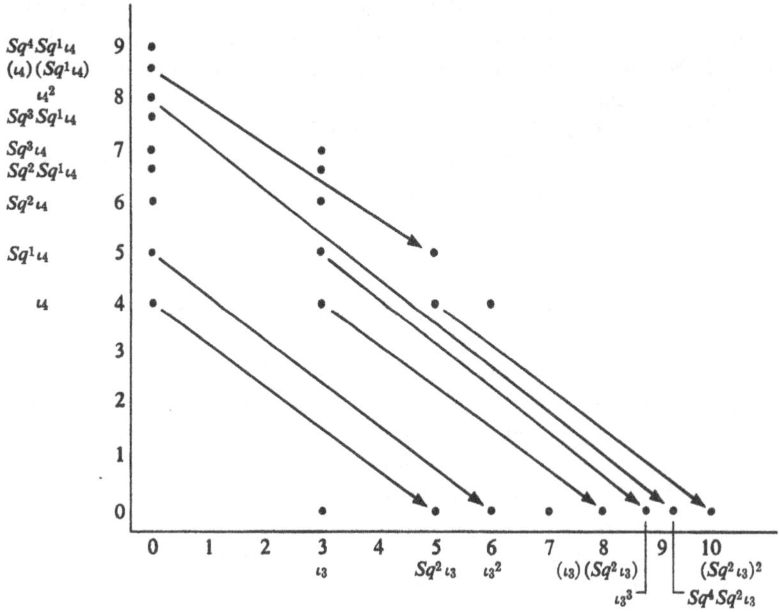

Figure 2

a diagram of the spectral sequence for $F_1 \to X_1 \to B$. The cohomology of the base and fibre are written out along the horizontal and vertical axes, and product terms are written at the appropriate locations in the first quadrant. In Figure 2 the elements of E_2 are indicated only by small circles, since in the scheme of the diagram the name of each term may be inferred from its location. Only terms with total degree of at most 10 need be considered for our purposes.

The calculation of the spectral sequence in this range is indicated in Figure 2. We have $d_5(\iota_4) = \tau(\iota_4) = Sq^2\iota_3$ by construction. As a corollary,

$d_5(\iota_3 \otimes \iota_4) = (\iota_3)(Sq^2\iota_3)$, since each d_r is a derivation (and the products are consistent in this sense). We have $d_5(Sq^1\iota_4) = 0$, $d_6(Sq^1\iota_4) = \tau(Sq^1\iota_4) = Sq^1(\tau(\iota_4)) = Sq^3\iota_3 = (\iota_3)^2$. Several elements transgress to zero; for example,

$$\begin{aligned}\tau(Sq^2Sq^1\iota_4) &= Sq^2Sq^1(\tau(\iota_4))\\ &= Sq^2Sq^1Sq^2\iota_3\\ &= Sq^5\iota_3 + Sq^4Sq^1\iota_3\end{aligned}$$

using the Adem relations, and thus $\tau(Sq^2Sq^1\iota_4) = 0$. Also, $Sq^4Sq^1\iota_4$ transgresses to zero, since

$$\begin{aligned}\tau(Sq^4Sq^1\iota_4) &= Sq^4Sq^1Sq^2\iota_3\\ &= Sq^4(\iota_3)^2\\ &= (Sq^2\iota_3)^2\end{aligned}$$

which is zero in E_{10} because $(Sq^2\iota_3)^2 = d_5(Sq^2\iota_3 \otimes \iota_4)$. In this calculation we have obtained $Sq^4(\iota_3)^2$ by the Cartan formula. Another proof that $Sq^4Sq^1Sq^2\iota_3 = (Sq^2\iota_3)^2$ is obtained by using the Adem relations:

$$Sq^4Sq^1Sq^2 = Sq^4Sq^3 = Sq^5Sq^2,$$

from which the result follows.

We find that the following elements of total degree ≤ 9 survive to E_∞ of this spectral sequence:

$$\iota_3 \qquad Sq^2\iota_4 \qquad Sq^3\iota_4 \qquad Sq^2Sq^1\iota_4 \qquad Sq^3Sq^1\iota_4$$
$$Sq^4Sq^1\iota_4 \qquad \iota_3 \otimes Sq^2\iota_4$$

Recall that the groups $E^{s,t}$ for which $s + t = n$ form a composition series for $H^n(X_1)$. We have no problems with group extensions when we are working over Z_2. Therefore we can easily write a basis for $H^n(X_1)$, $n \leq 9$. We denote the elements of this basis by ι_3, A, B, C, D, E, and $\iota_3 A$, respectively. For example, $B \in H^7(X_1)$ is defined by $i^*B = Sq^3\iota_4 \in H^7(F_1)$.

Since $i^*(Sq^1A) = Sq^1(i^*A) = Sq^1Sq^2\iota_4 = Sq^3\iota_4 = i^*B$, and since

$$i^*\colon H^7(X_1) \to H^7(F_1)$$

is monomorphic, we have $Sq^1A = B$. Therefore we should kill A by forming the induced fibre space over X_1 by a map of X_1 into $K(Z_2,6)$ corresponding to A. We thus have the following diagram:

$$\begin{array}{ccccc} F_2 = K(Z_2,5) & \to & X_2 & & \\ & & \downarrow & & \\ F_1 = K(Z_2,4) & \to & X_1 & \xrightarrow{A} & K(Z_2,6) \\ & & \downarrow & & \\ & & B = K(Z,3) & \xrightarrow{Sq^2} & K(Z_2,5) \end{array}$$

Using the Serre spectral sequence of the new fibre space $F_2 \to X_2 \to X_1$, we can calculate $H^*(X_2)$ at least through dimension 8. This is indicated in Figure 3. Just as we have seen that $Sq^1 A = B$, we can show that $Sq^2 A = D$. Note also that $i^*(Sq^2 Sq^1(A)) = Sq^2 Sq^1 Sq^2 \iota_4 = Sq^4 Sq^1 \iota_4 = i^*E$, while $i^*(Sq^3(A)) = Sq^3 Sq^2 \iota_4 = 0$. Since $\iota_3 A = d_6(\iota_3 \otimes \iota_5)$, this implies that $\tau(Sq^2 Sq^1 \iota_5) = E$ while $\tau(Sq^3 \iota_5) = 0$.

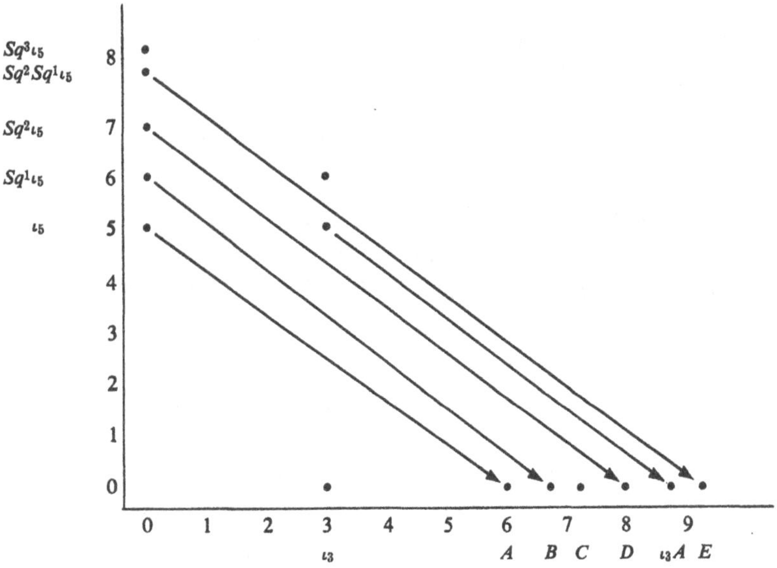

Figure 3

Thus the only survivors to E_∞ in total degree ≤ 8 are ι_3, C, and $Sq^3\iota_5$. It follows that $H^k(X_2)$ is generated by ι_3 in dimension 3, a class $C' = p^*C$ in dimension 7, and a class S in dimension 8 such that $i^*S = Sq^3\iota_5$. From our calculations, it follows that $\pi_4(S^3) = Z_2$ and $\pi_5(S^3) = Z_2$. To determine $\pi_6(S^3)$, we need only determine which Bockstein operator is non-zero on C'. Now C was defined by $i^*C = Sq^2 Sq^1 \iota_4 \in H^7(F_1)$, and $Sq^1 C = D = \tau(Sq^2 \iota_4) \in H^8(X_1)$. Thus the Bockstein lemma implies that

$$i^* d_2 C' = d_1 Sq^2 \iota_5 = Sq^3 \iota_5 = i^* S$$

and so the second Bockstein d_2 is non-zero. It is easy to verify that we can kill $H^7(X_2)$ by means of a map of X_2 into $K(Z_4, 7)$ corresponding to a cohomology class with Z_4 coefficients which reduces to C' (mod 2). Thus we have $\pi_6(S^3) = Z_4$.

We will not try to push these calculations any further.

Recall that, by the suspension theorem, the suspension homomorphism $E: \pi_i(S^n) \to \pi_{i+1}(S^{n+1})$ is isomorphic for $i < 2n-1$ and epimorphic for $i = 2n-1$. Thus $\pi_4(S^3)$ is isomorphic to $\pi_{n+1}(S^n)$ for any $n \geq 3$, that is, $\pi_4(S^3)$ is stable. Although $\pi_5(S^3) \to \pi_6(S^4)$ is not an isomorphism *a priori*, we can deduce that it is isomorphic from the fact that it is epimorphic and the fact that $\pi_5(S^3) = Z_2 = \pi_6(S^4)$. Thus $\pi_5(S^3)$ is also stable, not by definition but, as it were, by accident. On the other hand, $\pi_6(S^3) = Z_4$ is not stable, since we have seen that $\pi_{n+3}(S^n) = Z_8$ for $n \geq 5$.

EXERCISES

1. Let $M_n = S^n \cup_{2\iota} e^{n+1}$, where $2\iota: S^n \to S^n$ has degree 2. (Thus M_n is the $(n-1)$-fold suspension of the projective plane.) Calculate $\pi_{n+k}(M_n)$ for large n, small k.

2. Calculate $\pi_k(S^4)$ for $k \leq 7$ by the methods of the Appendix.

REFERENCES

The suspension theorem
1. R. Bott [1].
2. H. Freudenthal [1].

Generalizations and the EHP-sequence
1. G. W. Whitehead [2,4].

Composition-theoretic methods of calculation
1. M. G. Barratt [2,3].
2. H. Toda [1].

The second stable homotopy group of spheres
1. G. W. Whitehead [3].

CHAPTER 13

n-TYPE AND POSTNIKOV SYSTEMS

To work inductively in homotopy theory, we would like to have a sequence of invariants of homotopy type such that the nth invariant represents the given complex "through dimension n." Unfortunately the obvious candidate, namely, the n-skeleton of a complex, is not a homotopy invariant. In fact it is easy to find two complexes which have the same homotopy type but in which the n-skeletons do not have the same homotopy type.

A simple example is obtained by giving two different cell structures for the sphere. Let K denote S^n considered as a complex having one 0-cell and one n-cell where the n-cell is attached by the constant map on its boundary. Let L denote S^n obtained from S^{n-1} by attaching two n-cells (as the "hemispheres" of S^n attached to the "equator"). Then K and L are even homeomorphic as topological spaces, but the $(n-1)$-skeletons do not have the same homotopy type, since K^{n-1} is a point, whereas $L^{n-1} = S^{n-1}$.

We can circumvent this difficulty to some extent by using J. H. C. Whitehead's theory of n-types. First we recall (without proof) a useful technical result, the cellular-approximation theorem, also due to Whitehead.

We will use the letters K,L to denote CW complexes and X,Y to denote arbitrary topological spaces.

CELLULAR APPROXIMATION

Definition

A map $f: K \to L$ is *cellular* if $f(K^n) \subset L^n$ for every n, where K^n denotes the n-skeleton of K. A homotopy $F: K \times I \to L$ between cellular maps is cellular if $F(K^n \times I) \subset L^{n+1}$ for every n.

Theorem 1 (cellular-approximation theorem)

Let K_0 be a subcomplex of K. Let $f: K \to L$ be a map such that $f \mid K_0$ is cellular. Then there exists a cellular map $g: K \to L$ such that $g \sim f$ rel K_0.

Corollary 1

Let $f: K \to L$. Then there exists a cellular map $g: K \to L$ such that $g \sim f$.

Corollary 2

Let $f_0, f_1: K \to L$ be cellular maps. Suppose $f_0 \sim f_1$. Then there exists a cellular homotopy $F: K \times I \to L$ between f_0 and f_1.

n-TYPE

We will define two different notions, n-type and n-homotopy type. The reader is warned not to confuse them.

Definition

Two maps f,g of X into Y are *n-homotopic* if, for every complex K of dimension at most n and for every map φ of K into X, the compositions $f\varphi, g\varphi: K \to Y$ are homotopic.

It is clear that this defines an equivalence relation among maps of X into Y (for each n).

Proposition 1

Let f,g be maps of a complex K into a space X. Then f and g are n-homotopic if and only if the restrictions of f and g to K^n are homotopic.

PROOF: If f,g are n-homotopic, then by definition their compositions with the inclusion map of K^n into K are homotopic, but these are precisely the restrictions as stated. To prove the converse, it is enough to know that any map of an n-dimensional complex into K is homotopic to a map of that complex into K^n. This follows immediately from the cellular-approximation theorem.

Definition

Two spaces X, Y have the same *n-homotopy type* if there exists a map $f: X \to Y$ and a map $g: Y \to X$ such that the compositions fg and gf are n-homotopic to the respective identity maps. Then we say that (f,g) is an n-homotopy equivalence and that g is an n-homotopy inverse of f.

Having the same n-homotopy type is clearly an equivalence relation.

We now come to the definition which will be essential in the sequel.

Definition

Two complexes K,L have the same *n-type* if their n-skeletons K^n, L^n have the same $(n-1)$-homotopy type.

That is, K and L have the same n-type if there are maps between K^n and L^n such that the compositions are $(n-1)$-homotopic to the identity maps of K^n and L^n.

If we allow n to take the value ∞, it is with the understanding that "∞-homotopic" means homotopic, "K^∞" means K, and so forth. This is a convention, not a theorem; it is not true, for example, that two maps $f,g: X \to Y$ must be homotopic if $f\varphi$ and $g\varphi$ are homotopic for every map φ of a complex into X (see Exercise 1). However, the propositions that follow will hold true when n is allowed to take the value ∞.

We use cellular approximation again in the next result.

Proposition 2

If K and L have the same n-type, then they have the same m-type for every $m < n$.

PROOF: We are given that there exist maps $f: K^n \to L^n$ and $g: L^n \to K^n$ such that the compositions are $(n-1)$-homotopic to the respective identity maps. We may assume that f, g, and the homotopies are cellular. Then the restrictions of f and g to the m-skeletons provide an $(m-1)$-homotopy equivalence of the m-skeletons, which proves the proposition.

This proposition is valid with $n = \infty$; that is, if K and L have the same homotopy type, then they have the same m-type for every finite m. Thus the n-type is a homotopy invariant of the complex.

Recall the two cell structures for S^n given at the beginning of this chapter. It is easy to verify in this example that the m-type is invariant. For instance, if $n = 3$, then $K^2 = e^0$ and $L^2 = S^2$ have the same 1-homotopy type (since any map of a 1-dimensional complex into S^2 is null-homotopic); so K and L have the same 2-type.

It is not hard to see that the 1-type of a complex is essentially a measure of the number of connected components. Whitehead showed that two complexes have the same 2-type if and only if they have the same fundamental group. The main subject of this chapter will be Postnikov systems, which provide a characterization of n-type for larger values of n.

Recall that $[X, Y]$ denotes the set of homotopy classes of maps of X into Y.

Lemma 1

Let (f,g) be an n-homotopy equivalence of X and Y and let K be a complex of dimension n. Then the induced transformation $f_{\#}: [K,X] \to [K, Y]$ is a set isomorphism.

This is immediate; for the compositions $f_{\#}g_{\#}$ and $g_{\#}f_{\#}$ are the identity transformations.

Theorem 2

Suppose L_1 and L_2 have the same n-type. Let $\dim K \leq n-1$. Then the sets $[K,L_1]$ and $[K,L_2]$ are in one-to-one correspondence.

PROOF: By Theorem 1, the inclusions $L_i^n \to L_i$ for $i = 1,2$ induce one-to-one correspondences $[K,L_i^n] \to [K,L_i]$. By hypothesis, L_1^n and L_2^n have the same $(n-1)$-homotopy type. The result follows from Lemma 1.

Corollary 3

If K and L have the same n-type, then the homotopy groups $\pi_i(K)$ and $\pi_i(L)$ are isomorphic for all $i < n$.

The converse does not hold, but we have the following theorem of Whitehead, which we state without proof.

Theorem 3

Suppose there is a map $f: K^n \to L^n$ which induces isomorphisms on the homotopy groups $\pi_i(K^n) \approx \pi_i(L^n)$ for all $i < n$. Then K and L have the same n-type.

POSTNIKOV SYSTEMS

Now suppose that L is an $(n-1)$-connected complex, and let $\pi = \pi_n(L)$. Then L and $K(\pi,n)$ have the same homotopy groups through dimension n.

Proposition 3

L and $K(\pi,n)$ have the same $(n+1)$-type.

PROOF: We need a map between these two spaces which induces the isomorphism of π_n. Such a map $f: L \to K(\pi,n)$ is given, using the results of Chapter 1, by the fundamental class $\iota_n \in H^n(L;\pi)$. We leave the details to the reader (see Exercise 3).

A natural line of inquiry would be to attempt to modify $K(\pi,n)$ in such a way as to obtain a space having the same $(n+2)$-type as the original complex L. This leads to the development of Postnikov systems, the main idea of this chapter.

Definition

Let X be an $(n-1)$-connected complex, with $n \geq 2$ (thus X is simply connected). A diagram of the form of Figure 1 is called a *Postnikov system* for X if it satisfies the following conditions:

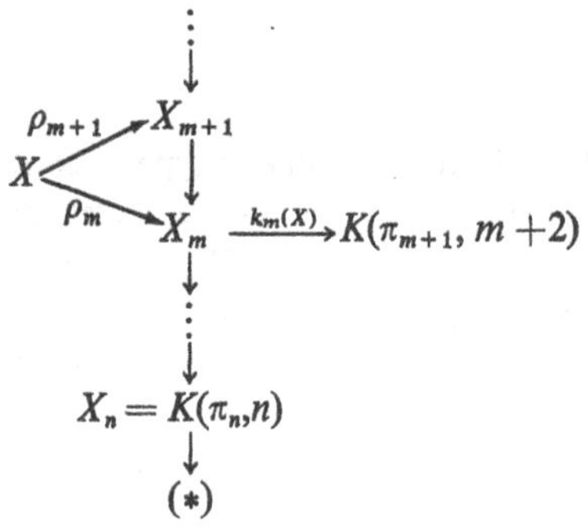

Figure 1

1. Each X_m $(m \geq n)$ has the same $(m+1)$-type as X, and there is a map $\rho_m\colon X \to X_m$ inducing the isomorphisms $\pi_i(X) \approx \pi_i(X_m)$ for all $i \leq m$
2. All higher homotopy groups of X_m are trivial, that is, $\pi_i(X_m)$ is trivial for $i > m$
3. X_{m+1} is the induced fibre space over X_m induced by $k_m(X)$ from the standard contractible fibre space
4. The diagram is homotopy-commutative

In Figure 1, π_i denotes $\pi_i(X)$; $k_m(X)$ denotes an appropriate cohomology class in $H^{m+2}(X_m;\pi_{m+1})$ or the corresponding homotopy class of maps; and $(*)$ denotes a space consisting of only one point.

We have already remarked that X and $K(\pi_n,n)$ have the same $(n+1)$-type. The construction of a Postnikov system is an extension of this result.

EXISTENCE OF POSTNIKOV SYSTEMS

We will now give an inductive construction to prove that for any simply connected space X there is a Postnikov system of arbitrarily high order.

The induction begins with the above observation that X and $X_n = K(\pi_n, n)$ have the same $(n+1)$-type.

Suppose that the following have been obtained through a certain dimension N: the finite Postnikov system consisting of the X_m, ρ_m, etc., for all $m \leq N$; and spaces $Y_n, \ldots, Y_N$ with $Y_n \subset Y_{n+1} \subset \cdots \subset Y_{N-1} \subset Y_N$ such that the maps ρ_m $(n \leq m \leq N)$ are factored as $\rho_m = r_m \circ i_m$:

$$\rho_m\colon X \xrightarrow{i_m} Y_m \xrightarrow{r_m} X_m$$

where $i_m\colon X \to Y_m$ is an inclusion map and $r_m\colon Y_m \to X_m$ is a fibre map such that r_m induces isomorphisms $\pi_i(Y_m) \approx \pi_i(X_m)$ for all $i \leq m$ and that i_m is a homotopy equivalence. We thus assume the existence of a diagram, Figure 2, where each r_m is a fibre map, all the inclusions are homotopy equivalences, and $\pi_i(X_m) = 0$ for $i > m$.

$$\begin{array}{ccccc}
Y_N & \xrightarrow{r_N} & X_N & & \\
\cup & & \downarrow & & \\
Y_{N-1} & \xrightarrow{r_{N-1}} & X_{N-1} & \xrightarrow{k_{N-1}(X)} & K(\pi_N, N+1) \\
\cup & & \downarrow & & \\
\vdots & & \vdots & & \\
\cup & & \downarrow & & \\
X = Y_n & \xrightarrow{r_n} & X_n & \xrightarrow{k_n(X)} & K(\pi_{n+1}, n+2)
\end{array}$$

Figure 2

We will now show how to obtain the spaces X_{N+1} and Y_{N+1} and the maps i_{N+1} and r_{N+1} which satisfy the same requirements.

Let F_m denote the fibre of the fibre map $r_m\colon Y_m \to X_m$. It is clear from the homotopy exact sequence of this fibre map that

$$\pi_i(F_m) = 0 \qquad i \leq m$$

$$\pi_i(F_m) \approx \pi_i(Y_m) \qquad i > m$$

with the isomorphism induced by the inclusion $j_m\colon F_m \to Y_m$. Thus F_m is m-connected and $\pi_{m+1}(F_m) \approx \pi_{m+1}(Y_m) \approx \pi_{m+1}(X)$, which we write simply as π_{m+1}.

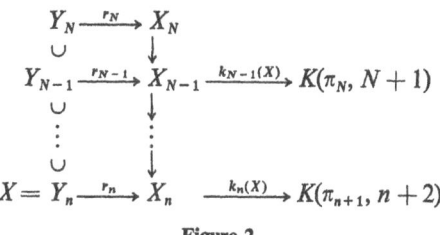

Recall that the characteristic class of this fibre space is an element of $H^{m+2}(X_m; \pi_{m+1})$, defined as the transgression of the fundamental class of F_m. We denote this characteristic class for $m = N$ by $k_N(X)$, and we take a map $X_N \to K(\pi_{N+1}, N+2)$ from the corresponding homotopy class and construct a fibre space over X_N as the induced fibre space from the standard contractible fibre space. The total space of this fibre space will be X_{N+1}. It is clear from the homotopy exact sequence that X_{N+1} has the required homotopy groups. Consider the diagram of Figure 3, where the maps g and s are yet to be constructed. Now $k_N(X)$ is in the image of the

$$\begin{array}{ccccccc} K = K(\pi_{N+1}, N+1) & & \longrightarrow & X_{N+1} & & \\ \uparrow s & & \nearrow g & \downarrow p & & \\ F_N \xrightarrow{j_N} & Y_N & \xrightarrow{r_N} & X_N & \xrightarrow{k_N(X)} & K(\pi_{N+1}, N+2) \end{array}$$

Figure 3

transgression, and so $r_N^*(k_N(X)) = 0$, and therefore the composition $k_N r_N\colon Y_N \to K(\pi_{n+1}, N+2)$ is null-homotopic. Therefore the map r_N has a lifting $g\colon Y_N \to X_{N+1}$. We will prove in a moment that g may be chosen so that it induces the isomorphism $\pi_{N+1}(F_N) \approx \pi_{N+1}(X_{N+1}) (= \pi_{N+1})$. Granting this, we convert g to a fibre map $r_{N+1}\colon Y_{N+1} \to X_{N+1}$ by the procedure of Proposition 1 of Chapter 9; then Y_N is a subspace of Y_{N+1} in a natural way and such that the inclusion is a homotopy equivalence; and $i_{N+1}\colon X \to Y_{N+1}$ will be the composite of this inclusion with i_N. Thus the inductive step of the construction of the Postnikov system is completed, after we have proved that g can be chosen as asserted.

From the null-homotopy of $k_N r_N\colon Y_N \to K(\pi_{N+1}, N+2)$ to the constant map, we may construct a lifting $g'\colon Y_N \to X_{N+1}$ such that $pg' = r_N$ (Proposition 1 of Chapter 11). Moreover, this g' maps fibres into fibres, since $g'(r_N^{-1}(x)) \subset p^{-1}(x)$ $(x \in X)$; we thus obtain a map

$$s'\colon F_N \to K = K(\pi_{N+1}, N+1)$$

by restriction. Now s' defines a cohomology class in $H^{N+1}(F_N; \pi_{N+1})$ (which we denote also by s'), and the fundamental class of K transgresses to $k_N(X)$ by construction; so s' transgresses to $k_N(X)$ by naturality. But the fundamental class ι of F_N also transgresses to $k_N(X)$; in fact this was the definition of $k_N(X)$. Since $\tau(s') = \tau(\iota)$, we have $s' = \iota + j_N^*(\alpha)$ for some $\alpha \in H^{N+1}(Y_N; \pi_{N+1})$, by the Serre cohomology exact sequence.

Now the standard contractible fibre space of paths over $K(\pi_{N+1}, N+2)$ may be taken to be a principal fibre space (see Chapter 11), and then the induced fibre space $K \to X_{N+1} \to X_N$ is also a principal fibre space, with an action $\varphi\colon X_{N+1} \times K \to X_{N+1}$. We define the map $g\colon Y_N \to X_{N+1}$ by the composition

$$g\colon Y_N \xrightarrow{(g', -\alpha)} X_{N+1} \times K \xrightarrow{\varphi} X_{N+1}$$

where $-\alpha$ is a well-defined element of the group $H^{N+1}(Y_N; \pi_{N+1})$ and hence represents a homotopy class of maps $Y_N \to K$. Then $pg = pg' = r_N$ (by condition (1) for a principal fibre space), and we can define a map $s: F_N \to K$ by restriction of g. It follows from the definitions of s and s' and that of g that $s = \varphi \circ (s', -j_N^*(\alpha))$. But the restriction of φ to $K \times K$ is loop multiplication in $K = \Omega K(\pi_{N+1}, N+2)$, which corresponds to addition in $H^{N+1}(F_N; \pi_{N+1})$; and so we have, in cohomology, the relation $s = s' - j_N^*(\alpha)$. Thus $s = \iota$.

Since $s = \iota$, the map s induces an isomorphism on the $(N+1)$th homotopy group. But it is clear that the inclusion maps of the fibres also induce isomorphisms on this group. From the commutativity of the "square" in Figure 3, it follows that g also induces an isomorphism on π_{N+1}, which was to be proved.

This completes the construction of the Postnikov system for X. By Milnor's result (see Chapter 9) we may assume the spaces X_m are nice.

NATURALITY AND UNIQUENESS

It will be convenient to have a uniqueness theorem for Postnikov systems, asserting that any two such systems for a given space X are in some sense equivalent. It will also be useful to have the "functorial" property that a map from a space X to a space Y gives rise to a "map" from the Postnikov system of X to that of Y. These properties are consequences of the following theorem.

Theorem 4

Let X, Y be simply connected complexes, and let f be a map from X to Y. Let Postnikov systems be given for X and for Y. Then there exists a family of maps $\{f_m: X_m \to Y_m\}$ with the following properties:

1. $f_m^*(k_m(Y)) = f_\# k_m(X) \in H^{m+2}(X_m; \pi_{m+1}(Y))$
2. The diagram

$$\begin{array}{ccc} X_{m+1} & \xrightarrow{f_{m+1}} & Y_{m+1} \\ \downarrow & & \downarrow \\ X_m & \xrightarrow{f_m} & Y_m \end{array}$$

is commutative, and the diagram

$$\begin{array}{ccc} X & \xrightarrow{f} & Y \\ {\scriptstyle p_m}\downarrow & & \downarrow{\scriptstyle p_m'} \\ X_m & \xrightarrow{f_m} & Y_m \end{array}$$

is homotopy-commutative

3. If f and g are homotopic maps from X to Y, then f_m and g_m are homotopic (for every m)

In (1) we mean by $f_{\#}$ the coefficient homomorphism induced by the homomorphism $f_{\#}: \pi_{m+1}(X) \rightarrow \pi_{m+1}(Y)$.

As a corollary, X_m is determined by X, up to homotopy type; for we may take $Y = X$ and apply (3) to obtain a homotopy equivalence between the two versions of X_m.

We omit to prove this theorem; the reader now has all the necessary tools available.

DISCUSSION

If X is a complex, the n-skeleton X^n has the same homology (or cohomology) as X through dimension $n-1$ and has no homology above dimension n. Evidently the term X_n in a Postnikov system for X plays an analogous role for homotopy, but of course the situation is much more complicated and geometrically much less natural.

If X has only finitely many non-trivial homotopy groups, so that $\pi_i(X) = 0$ for all $i > N$ (for some N), then X_N has the same homotopy type as X and we can say that X has a finite Postnikov system.

If K is a complex of dimension n (or less), then we can identify the sets of homotopy classes $[K,X] = [K,X_n]$, since X and X_n have the same $(n+1)$-type. This simplification of the task of computing $[K,X]$ will be pursued further in Chapter 14.

In a certain sense, the Postnikov system provides a complete system of invariants for n-type for every n $(1 \leq n \leq \infty)$. For example, let X be $(n-1)$-connected. Then the $(n+1)$-type of X is that of $K(\pi_n(X),n)$, so that π_n is a complete set of invariants for $(n+1)$-type. (We have proved this for $n > 1$; it holds for $n = 1$, but the details are more complicated.) The $(n+2)$-type of X is that of X_{n+1}, and so it is completely determined by π_n, π_{n+1}, and k_n. Here $k_n \in H^{n+2}(\pi_n,n; \pi_{n+1})$. In general, the $(n+q+2)$-type of X is that of X_{n+q+1} and thus is determined by the groups $\pi_n,\ldots,\pi_{n+q+1}$ and the cohomology classes $k_n,\ldots,k_{n+q}$. In our formulation, it is not quite precise to call the k_n's invariants. Indeed, one may have fibre spaces of the same homotopy type induced from the same contractible fibre space by nonhomotopic maps. Nevertheless the k_n's are called the *k-invariants* of the space X.

Postnikov systems were indeed discovered (and called "natural systems") by Postnikov. Theorem 4 precisely expresses the manner in which the Postnikov system is a natural system; this theorem was proved by D. W. Kahn.

EXERCISES

1. Give an example of spaces X, Y and maps $f,g\colon X \to Y$ such that $f\varphi \sim g\varphi$ for any map φ of any complex K into X but with f,g not homotopic.
2. Give a counterexample to the converse of Corollary 3.
3. Let L be $(n-1)$-connected and let $\pi = \pi_n(L)$. Verify from the definitions that the fundamental class ι of L gives a homotopy class of maps $L \to K(\pi,n)$ which induce the isomorphism of π_n.
4. Show that the homotopy classes $[K,X_n]$ are in one-to-one correspondence with the set $i^{\#}([K^{n+1},X]) \subset [K^n,X]$. Here K is a complex, $i\colon K^n \subset K^{n+1}$ is the inclusion of the skeletons, X is a simply connected space, and X_n has the $(n+1)$-type of X but with $\pi_i(X_n)$ trivial for $i > n$.

   Show that the above correspondence is natural, in the obvious sense, if X_n is taken to be the nth term in a Postnikov system for X.

REFERENCES

Cellular approximation and n-type

1. P. J. Hilton [1].
2. J. H. C. Whitehead [1].

Postnikov systems

1. D. W. Kahn [1].
2. M. M. Postnikov [1,2,3,4].
3. E. H. Spanier [1].

CHAPTER 14

MAPPING SEQUENCES AND HOMOTOPY CLASSIFICATION

We will present two constructions of a spectral sequence for $[K,X]$. One is based on the Postnikov system for X; the other, on the cell structure of K.

THE FIBRE MAPPING SEQUENCE

Let $(E,p,B;F)$ be a fibre space in the sense of Hurewicz. If we take the inclusion $j\colon F \to E$ of the fibre and convert this map to a fibre map (Proposition 1 of Chapter 9) and then repeat the process on the inclusion of the new fibre and continue *ad infinitum*, we obtain what we will call the *fibre mapping sequence.*

The interest of the mapping sequence lies in the fact that it has a kind of periodicity. Precisely, when we convert j to a fibre map $j'\colon F' \to E$, the fibre G is a space of the same homotopy type as ΩB. Granting this, then similarly the inclusion $k\colon G \to F'$ gives a fibre map $k'\colon G' \to F'$ with a fibre H of the same homotopy type as ΩE. Moreover, it turns out that the inclusion of H in G' is, except for sign, homotopically equivalent to the map $\Omega p\colon \Omega E \to \Omega B$. (For the proofs of these facts, see the Appendix to this chapter.)

We thus obtain a sequence which may be written

$$\cdots \to \Omega^2 B \to \Omega F \to \Omega E \to \Omega B \to F \to E \to B$$

where we have put F for F', ΩB for G', etc. Each map in the sequence is a Hurewicz fibre map. Therefore, if K is any complex, the homotopy classes of maps of K into the terms of this sequence give an *exact sequence*

$$\cdots \to [K,\Omega^2 B] \to [K,\Omega F] \to [K,\Omega E] \to [K,\Omega B] \to [K,F] \to [K,E] \to [K,B]$$

Here the last three terms are not necessarily groups, but they are at least sets with distinguished element, so that exactness makes sense.

We remark that if the original fibre space $F \to E \to B$ is induced by a map $B \to Y$ from the standard contractible fibre space over Y, then we can add a term at the end of the above exact sequence, so that it terminates

$$\cdots \to [K,F] \to [K,E] \to [K,B] \to [K,Y]$$

In general the set of homotopy classes $[K,X]$ has a natural group structure if X is a space of loops. Thus we can make the following remark.

Proposition 1

$[K,X]$ has a natural group structure if the space X is $(n-1)$-connected and the dimension of the complex K is at most $2n-2$.

This follows from the fact that X and ΩSX have the same $(2n-1)$-type, by the suspension theorem.

In the sequel we always assume $[K,X]$ has a natural group structure. Other sufficient conditions include K being a suspension or X a loop space (see Chapter 15).

We will make implicit use of some facts about the behavior of cohomology classes and operations under the action of the looping functor Ω. Consider a cohomology class $\theta \in H^{n+k}(\pi,n; G)$ and let θ also denote a corresponding map of $K(\pi,n)$ into $K(G, n+k)$. Then $\Omega\theta$ is a map of $K(\pi, n-1)$ into $K(G, n+k-1)$, which corresponds to a cohomology class in $H^{n+k-1}(\pi, n-1; G)$. In this way we obtain a transformation

$$H^{n+k}(\pi,n; G) \to H^{n+k-1}(\pi, n-1; G)$$

which is called the *cohomology suspension* and written $^1(\ )$, so that the class which we denoted by $\Omega\theta$ above is written $^1\theta$. It is clear that the above transformation is an isomorphism in the range of Serre's exact sequence. We have commutative diagrams

$$\begin{array}{ccc} K(Z, n-1) & \xrightarrow{Sq^i} & K(Z_2, n+i-1) \\ \| & & \| \\ \Omega K(Z,n) & \xrightarrow{^1Sq^i} & \Omega K(Z_2, n+i) \end{array}$$

(and similarly with Z_2 replacing Z), and in this sense we have $^1(Sq^i) = Sq^i$. One expresses this fact by saying that the Sq^i are stable operations.

MAPPINGS OF LOW-DIMENSIONAL COMPLEXES INTO A SPHERE

In Chapter 12 we computed some homotopy groups of spheres by a construction which may be called the "mod 2 Postnikov system" for S^n. In fact, up to dimension $n+2$, this is a Postnikov system, with no apologies about mod 2, according to the following results.

Proposition 2

Let p be an odd prime. Then the least integer n such that $\pi_n(S^3)$ has a non-zero element of order p is $n = 2p$.

For an indication of the proof, see Exercise 1.

Corollary 1

$\pi_{n+k}(S^n)$ contains no elements of odd order for $k \leq 2$.

PROOF: From the proposition, $\pi_n(S^3)$ has no elements of odd order for $n < 6$. By the suspension theorem, $\pi_{n+1}(S^n) \approx \pi_4(S^3)$ and this group must be Z_2, since the 2-component is Z_2 and there is no odd torsion. Also, $\pi_{n+2}(S^n) \approx \pi_6(S^4)$, and this is the homomorphic image of $\pi_5(S^3) = Z_2$; so $\pi_{n+2}(S^n)$ contains no elements of odd order.

Thus, through dimension $n+2$, the Postnikov system of S^n (for n large) is exactly the mod 2 system previously constructed.

We will use this Postnikov system and the fibre mapping sequence of certain maps to study the classification of maps of a complex K into S^n.

Suppose first that K has dimension n. Then $[K,S^n] = [K,K(Z,n)]$ since $K(Z,n)$ and S^n have the same $(n+1)$-type. Since $[K,K(Z,n)]$ is in 1-1 correspondence with $H^n(K;Z)$, we recover the classical Hopf classification theorem for $[K,S^n]$.

Suppose next that K has dimension $n+1$. We have $[K,S^n] = [K,X_{n+1}]$ where X_{n+1} is the appropriate stage in the Postnikov system for S^n. Consider the diagram below where the map $\Omega B \to F_1$ is essentially $\Omega(Sq^2) = Sq^2$.

$$\begin{array}{ccc} \Omega B = K(Z, n-1) \longrightarrow F_1 = K(Z_2, n+1) \xrightarrow{i} & X_{n+1} & \\ & \downarrow{\scriptstyle p} & \\ & B = K(Z,n) & \xrightarrow{Sq^2} K(Z_2, n+2) \end{array}$$

Mapping K into the successive spaces in the diagram, we obtain an exact sequence

$$[K,\Omega B] \xrightarrow{f} [K,F_1] \xrightarrow{g} [K,X_{n+1}] \xrightarrow{h} [K,B] \xrightarrow{k} [K, K(Z_2, n+2)]$$

in which each term is a group. (We recall that $K(\pi,n)$ can be thought of as $\Omega K(\pi, n+1)$ and $[K,K(\pi,n)] = H^n(K;\pi)$.) Thus we know $[K,X_{n+1}]$ up to a group extension, since we have a short exact sequence

$$0 \to \operatorname{coker} f \to [K, X_{n+1}] \to \ker k \to 0$$

where $\operatorname{coker} f = H^{n+1}(K; Z_2)/Sq^2 H^{n-1}(K; Z)$ and where $\ker k$ is the kernel of $Sq^2 \colon H^n(K;Z) \to H^{n+2}(K;Z_2)$.

We will push this method one step further and consider $[K,S^n]$ where K has dimension $n+2$. Here $[K,S^n] = [K,X_{n+2}]$ where X_{n+2} is the next stage in the Postnikov system for S^n. By converting certain maps to fibre maps according to the method described at the beginning of the chapter, we can expand the Postnikov system (up to X_{n+2}) into the diagram of Figure 1.

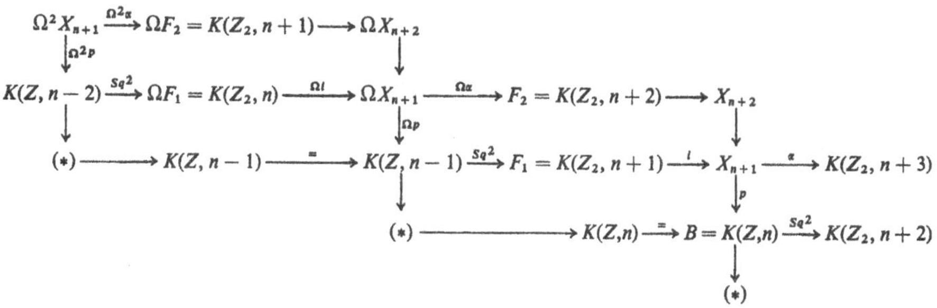

Figure 1 Diagram for analysis of $[K, S^n] = [K, X_{n+2}]$ ($\dim K = n + 2$)

Here each symbol $(*)$ denotes a one-point space, and we have used the remark that $^1(Sq^2) = Sq^2$. The map α was defined in Chapter 12 and satisfies the relation $i^*\alpha = Sq^2(\iota_{n+1}) \in H^{n+3}(F_1; Z_2)$.

To study $[K, X_{n+2}]$, we map K into the following portion of Figure 1:

$$\begin{array}{ccc} \Omega X_{n+1} \xrightarrow{\Omega\alpha} F_2 \to & X_{n+2} & \\ & \downarrow & \\ & X_{n+1} & \xrightarrow{\alpha} K(Z_2, n+3) \end{array}$$

and obtain a short exact sequence

$$0 \to \operatorname{coker}(\Omega\alpha)_\# \to [K, X_{n+2}] \to \ker \alpha_\# \to 0$$

Now $\alpha_\# \colon [K, X_{n+1}] \to [K, K(Z_2, n+3)]$ is the zero homomorphism, since $[K, K(Z_2, n+3)]$ is classified by $H^{n+3}(K; Z_2)$, which vanishes by the assumption on the dimension of K. Thus $\ker \alpha_\#$ is the entire group $[K, X_{n+1}]$, which we have described previously.

It is much more complicated to describe $\operatorname{coker}(\Omega\alpha)_\#$. Let A denote the kernel of $Sq^2 \colon H^{n-1}(K;Z) \to H^{n+1}(K; Z_2)$ and let C denote the cokernel of

$Sq^2: H^n(K; Z_2) \to H^{n+2}(K; Z_2)$. (The reader should identify these cohomology groups with the appropriate locations on Figure 1.) We can define a transformation

$$d_3 = (\Omega\alpha)(\Omega p)^{-1}: A \to C$$

In terms of the figure, we use the portion

$$\begin{array}{ccccc} \Omega F_1 = K(Z_2,n) & \longrightarrow & \Omega X_{n+1} & \longrightarrow & F_2 = K(Z_2, n+2) \\ & & \big\downarrow{\scriptstyle \Omega p} & & \\ K(Z, n-1) & \xrightarrow{=} & K(Z, n-1) & \longrightarrow & F_1 \end{array}$$

An element of A corresponds to a map of K into $K(Z, n-1)$ which can be lifted to a map into ΩX_{n+1} and thus continued to F_2, resulting in an element of $[K,F_2] = H^{n+2}(K;Z_2)$. This element is not well-defined there, since the lifting may vary, but is well-defined as an element of the quotient group C, since the lifting may only vary by the effect of a map of K into ΩF_1 and the composite $\Omega F_1 \to \Omega X_{n+1} \to F_2$ is the Sq^2 used to define C.

We claim that coker $(\Omega\alpha)_\#$ is isomorphic to coker d_3; we omit the proof, which is a tedious verification. Granting this, we obtain a composition series for $[K,X_{n+2}]$ with the following terms: (1) ker $Sq^2 \subset H^n(K;Z)$; (2) coker $Sq^2 = H^{n+1}(K;Z_2)/Sq^2H^{n-1}(K;Z)$; (3) coker d_3 as above. Of course (1) and (2) represent $[K,X_{n+1}]$.

THE SPECTRAL SEQUENCE FOR $[K,X]$

We hope that by now the reader can see what is coming. Abandoning any dimensional restriction on K and no longer assuming X a sphere, we can subsume the above cases into a general framework to obtain a spectral sequence for $[K,X]$.

We obtain the exact couple simply by mapping the complex K (of any dimension, now) into the diagram of Figure 1 (extended upwards and to the left according to a Postnikov system for X). We will denote the initial stage with the index 2. The D_2 terms are homotopy classes of maps of K into the terms of the Postnikov system for the space X and into the derived towers, i.e., into the first, third, and fifth columns of Figure 1. The E_2 terms come from maps of K into the fibres. The bigradation will come from the plane coordinates implicit in the diagram.

Explicitly, we have an exact couple

$$\begin{array}{ccc} D_2 & \xrightarrow{i_2} & D_2 \\ {\scriptstyle k_2}\nwarrow & & \swarrow{\scriptstyle j_2} \\ & E_2 & \end{array}$$

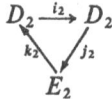

where $D_2^{-p,q} = [K,\Omega^p X_{q-p}]$ and $E_2^{-p,q} = [K,\Omega^p F_{q-p-n}]$. All the D's and E's are groups. It may be more convenient to use the natural equivalence $[K,\Omega Y] = [SK, Y]$ to obtain alternate expressions for these groups. For example,

$$\begin{aligned} E_2^{-p,q} = [K,\Omega^p F_{q-p-n}] &= [K,\Omega^p(K(\pi_m(X),m))] \qquad m = q-p \\ &= [S^p K, K(\pi_m(X),m)] \\ &= H^{q-p}(S^p K; \pi_{q-p}(X)) \end{aligned}$$

Thus $E_2^{-1,n+2} = [K,\Omega F_1] = [K,K(Z_2,n)] = H^{n+1}(SK; \pi_{n+1}(X))$.

The map i_2 is induced by the projections in the towers and has bidegree $(0,-1)$; thus, typically,

$$i_2\colon D_2^{-p,q+1} \to D_2^{-p,q}\colon [K,\Omega^p X_{q-p+1}] \to [K,\Omega^p X_{q-p}]$$

The map j_2 is induced by the k-invariants and has bidegree $(1,0)$:

$$j_2\colon D_2^{-p,q} \to E_2^{-p+1,q}\colon [K,\Omega^p X_{q-p}] \to [K,\Omega^{p-1} F_{q-p+1-n}]$$

The map k_2, of bidegree $(0,0)$, comes from the fibre inclusions:

$$k_2\colon E_2^{-p,q} \to D_2^{-p,q}\colon [K,\Omega^p F_{q-p-n}] \to [K,\Omega^p X_{q-p}]$$

Then of course $d_2 = j_2 k_2\colon E_2^{-p,q} \to E_2^{-p+1,q}$. This comes from "two steps sideways," for example, from $[K,\Omega F_1]$ to $[K,F_2]$. Observe that the d_2 is induced by a map of $K(\pi,n)$ spaces, so that it corresponds to a primary cohomology operation.

We leave it to the reader to trace the "path" of a higher differential on the diagram of Figure 1; d_r has bidegree $(1, r-2)$ for $r \geq 2$.

If K is finite-dimensional, each column of the diagram is finite and we can replace X by X_q for $q = \dim K$. We obtain a composition series for $[K,\Omega^r X] = [S^r K, X]$ where the successive quotients in the series are the terms $E_\infty^{-p,q}$ with $-p = r$. The filtration is defined by taking F_j to be the kernel of the map $[K,X] \to [K,X_j]$ induced by $\rho_j\colon X \to X_j$; we have $F_{j-1}/F_j = E_\infty^{0,j}$ and $0 \subset F_j \subset F_{j-1} \subset \cdots \subset F_0 = [K,X]$ (assuming X connected). The filtration is thus obtained by means of the Postnikov system for X.

The reader should reexamine the example $[K,S^n]$ where K has dimension $n+2$ to see how it fits into the general scheme.

THE SKELETON FILTRATION AND THE INCLUSION MAPPING SEQUENCE

The above theory can be dualized. Let K be a complex and L a subcomplex, with $i\colon L \to K$ the inclusion. Let j be the inclusion of K in $K \cup CL$.

Then, if X is a simply connected space,

$$[L,X] \xleftarrow{i^*} [K,X] \xleftarrow{j^*} [K \cup CL, X]$$

is an exact sequence (the verification is easy); in fact $L \to K \to K \cup CL$ is a "co-fibration." We have assumed that (K,L) is a pair of complexes so that $K \cup CL$ will have the same homotopy type as K/L.

We now consider $j: K \to K \cup CL$ and form $(K \cup CL) \cup CK$, which is the "co-fibre" of the inclusion map j. It is not hard to see that $(K \cup CL) \cup CK$ has the homotopy type of the suspension SL. The next step produces a complex of the same homotopy type as SK, and the map is homotopically equivalent to $Si: SL \to SK$. In this way we obtain the *inclusion mapping sequence* which may be written, up to homotopy equivalence, as

$$L \xrightarrow{i} K \xrightarrow{j} K/L \to SL \xrightarrow{Si} SK \xrightarrow{Sj} S(K/L) \to S^2L \to \cdots$$

Mapping the terms of the above sequence into a simply connected space X, we have the exact sequence

$$[L,X] \leftarrow [K,X] \leftarrow [K/L,X] \leftarrow [SL,X] \leftarrow [SK,X] \leftarrow [SK/SL,X] \leftarrow \cdots$$

where the first three terms are sets with distinguished point, the next three are groups, and all the rest are abelian groups.

We remark that it is possible to give a suitable definition of the cell structure of the complex SK so that $S(K/L) = SK/SL$ and, less obviously, so that $S^r(K^p) = (S^rK)^{p+r}$ (where K^p is the p-skeleton of K).

Fix p for the moment and note that K^p/K^{p-1} is a wedge of p-spheres. According to the preceding remarks about S^rK^p, we can identify

$$S^r(K^p/K^{p-1}) = S^rK^p/S^rK^{p-1} = (S^rK)^{p+r}/(S^rK)^{p+r-1}$$

which is a wedge of $(p+r)$-spheres, and we can then identify the maps of this wedge into X with a cochain group,

$$[(S^rK)^{p+r}/(S^rK)^{p+r-1}, X] \approx C^{p+r}(S^rK; \pi_{p+r}(X)) \approx C^p(K; \pi_{p+r}(X))$$

Thus if we take $K = K^p$, $L = K^{p-1}$, and let p vary, the resulting sequences fit together to form an exact couple, with a typical D_1 term given by $[S^rK^p, X]$ and a typical E_1 term by $[S^r(K^p/K^{p-1}), X] \approx C^p(K; \pi_{p+r}(X))$. The reader should draw an appropriate diagram. The differential

$$d_1: C^p(K;\pi_{p+r}(X)) \to C^{p+1}(K;\pi_{p+r}(X))$$

turns out to be the usual cohomology coboundary operator, up to sign. A typical E_2 term has form $H^p(K;\pi_{p+r}(X))$.

With a suitable bigrading chosen on E_1 and D_1, one finds that the E_2 term of the resulting spectral sequence is isomorphic to the E_2 term of the spectral sequence obtained from the exact couple of the fibre mapping sequence. By Exercise 2 of Chapter 13, the same holds for the D_2 terms. The filtration, instead of being based on the kernels of the maps $[K,X] \to [K,X_j]$, is based on the kernels of the maps $[K,X] \to [K^j,X]$. These filtrations coincide and we obtain the same spectral sequence for $[K,X]$.

DISCUSSION

We have given two constructions of the spectral sequence for $[K,X]$. Without precisely defining duality, we view these constructions as dual. We think of fibration and co-fibration as dual notions. The "building block" using fibrations is the Eilenberg-MacLane space; for co-fibrations, it is a wedge of spheres (having only one non-vanishing cohomology group).

The inclusion mapping sequence is often referred to as the Puppe sequence; it was first studied extensively by Barratt. The existence of the spectral sequence for $[K,X]$ is perhaps more obvious from this point of view. On the other hand, the Postnikov approach shows how the k-invariants of X and cohomology operations in K affect the result; namely, they determine d_2. Subsequently we will introduce secondary cohomology operations; the reader will then immediately be able to interpret the next differential.

APPENDIX: PROPERTIES OF THE FIBRE MAPPING SEQUENCE

Lemma 1A

Let $(E,p,B;F)$ be a Hurewicz fibre space. Let Z denote the subspace $\{(e,\beta): p(e) = \beta(0)\}$ of $E \times B^I$. Then there is a map $\lambda: Z \to E^I$ such that $\lambda(e,\beta)(0) = e$ and $p\lambda(e,\beta) = \beta$.

Here p denotes the map $E \to B$ and also the induced map $E^I \to B^I$. It will sometimes be convenient to use the notation $E * B^I$ for Z. Mapping spaces like E^I are always understood to have the compact-open topology.

PROOF: We have a map $Z \to E$ taking (e,β) into e; this projects to a map $Z \to B$ taking (e,β) into $p(e) = \beta(0)$. We have a homotopy of this latter map, taking $Z \times I$ into B by taking $(e,\beta; t)$ into $\beta(t)$. Applying the covering homotopy theorem, we obtain the map λ of the lemma.

Lemma 2A

The map $E^I \rightarrow E^I$ given by $\omega \rightarrow \lambda(\omega(0), p\omega)$ is homotopic to the identity map, and the homotopy is "vertical," that is, the image of any point remains in its fibre throughout the homotopy.

The proof may be left as an exercise.

Lemma 3A

Let $(E,p,B; F)$ be a Hurewicz fibre space. Convert the inclusion $j: F \rightarrow E$ into a fibre map as in Proposition 1 of Chapter 9. Let G be the fibre of $j': F' \rightarrow E$. Then G has the homotopy type of ΩB.

PROOF: Recall that $F' = \{(f,\omega): j(f) = \omega(0)\} \subset F \times E^I$ and that j' is defined by $j'(f,\omega) = \omega(1)$. Let e_0 denote the base point of the pair (E,F). Then $G = (j')^{-1}(e_0)$, so that

$$G = \{(f,\omega): \omega(0) = j(f),\ \omega(1) = e_0\} \subset F'$$

We have an obvious map $\varphi: G \rightarrow \Omega B: (f,\omega) \rightarrow p\omega$. We obtain a map $\psi: \Omega B \rightarrow G$ as follows. Given a path $\beta \in B^I$, let ε denote the inverse of the path $\lambda(e_0,\beta^{-1})$. Then put $\psi(\beta) = (\varepsilon(0),\varepsilon)$. This defines a map of ΩB into G, since $\varepsilon(1) = \lambda(e_0,\beta^{-1})(0) = e_0$.

We must prove that the compositions $\varphi\psi$ and $\psi\varphi$ are homotopic to the identity maps. Now $\varphi\psi(\beta) = \varphi(\varepsilon(0),\varepsilon) = p\varepsilon = \beta$, so that $\varphi\psi$ is actually the identity itself. In the other direction, $\psi\varphi(f,\omega) = \psi(p\omega) = (\gamma(0),\gamma)$ where $\gamma^{-1} = \lambda(e_0,(p\omega)^{-1})$. Now $(f,\omega) \in G$; so $\omega(1) = e_0$. Thus $\lambda(e_0,(p\omega)^{-1}) = \lambda(\omega^{-1}(0), p(\omega^{-1}))$; but the map $\omega \rightarrow \gamma^{-1} = \lambda(\omega^{-1}(0), p(\omega^{-1}))$ is homotopic to the map $\omega \rightarrow \omega^{-1}$, by the previous lemma. Thus the map $\omega \rightarrow \gamma$ is homotopic to the identity map $\omega \rightarrow \omega$. It follows easily that the map $\psi\varphi: G \rightarrow G: (f, \omega) \rightarrow (\gamma(0),\gamma)$ is homotopic to the map $(f,\omega) \rightarrow (\omega(0),\omega) = (f,\omega)$, i.e., to the identity. This proves the lemma.

Since the composition $\varphi\psi: \Omega B \rightarrow G \rightarrow \Omega B$ is the identity map, we have actually proved that ΩB may be identified with a deformation retract of G.

THE MAP FROM ΩE TO ΩB

Now consider the diagram below, where $(E,p,B; F)$ is the original Hurewicz fibre space. The inclusion $j: F \rightarrow E$ is converted to a fibre map j' with fibre G, and the inclusion $k: G \rightarrow F'$ is converted to a fibre map

k' with fibre H. The four inclusion maps indicated by vertical arrows are homotopy equivalences.

$$\begin{array}{ccccccc}
\Omega E & \xrightarrow{\Omega p} & \Omega B & & F & \xrightarrow{j} & E \xrightarrow{p} B \\
 & & \cap\downarrow & & \cap\downarrow & \nearrow j' & \\
\Omega E & & G & \xrightarrow{k} & F' & & \\
\cap\downarrow & & \cap\downarrow & \nearrow k' & & & \\
H & \xrightarrow{l} & G' & & & &
\end{array}$$

Lemma 4A

The maps l and Ωp are homotopically equivalent except for sign.

The proof requires no deep thinking; it consists principally of tracing the definitions through the diagram. We present it here for the curiosity value of the result that the homotopy equivalence is not the natural one but must be obtained by reversing all the paths in ΩE. Precisely, we will prove that the following diagram is homotopy-commutative where T is the homotopy equivalence which reverses the parametrization of every path and α, β, and γ are the retractions given by Proposition 1 of Chapter 9 (for β) and Lemma 3A (for α and γ). The need to use T in this lemma may be viewed as the source of an exasperating $(-1)^n$ which appears frequently in algebraic topology. In this book we usually work mod 2.

$$\begin{array}{ccc}
\Omega E & \xrightarrow{\Omega p} & \Omega B \\
T\uparrow & & \uparrow\gamma \\
\Omega E & & G \\
\alpha\uparrow & & \uparrow\beta \\
H & \xrightarrow{l} & G'
\end{array}$$

Let a typical "point" of H be (g,φ') where $g=(f,\omega)\in G$ and φ' is a path in F'. Since $(g,\varphi')\in H$, we have $\varphi'(0)=g$, $\varphi'(1)=(e_0,e_0^*)$ where e_0 is the base point in $F\subset E$ and e_0^* is the constant path there. Since $g\in G$, we have $\omega(0)=f$ and $\omega(1)=e_0$.

The two routes around the diagram give the following calculations for the image of (g,φ') in ΩB:

$$\gamma(\beta(l(g,\varphi')))=\gamma(\beta(g,\varphi'))=\gamma(g)=\gamma(f,\omega)=\Omega p(\omega)$$

$$\Omega p(T(\alpha(g,\varphi')))=\Omega p(T(\Omega j'(\varphi')))=T(\Omega p(\Omega j'(\varphi')))$$

where the last T denotes an operator on B. Now it is readily seen from the definitions that pj' takes a point $(e,\varepsilon)\in F'=F*E^I$ to the point $p(\varepsilon(1))\in B$. The path $\varphi'\in(F')^I$ is a continuous deformation of (f,ω) to the base point

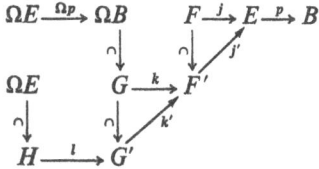

(e_0, e_0^*) of F'. Geometrically speaking, we take (f,ω) and move the point f along the fibre F to e_0, while the path ω is free (except that its initial point moves with f) until at the end of the deformation ω is collapsed to e_0^*. Then $\Omega p(\Omega j'(\varphi'))$ is the track in B of the endpoint of this path. If we write $\varphi'(t) = (f_t,\omega_t) \in F'$, then we must compare the path $t \to p(\omega(t))$ with the path $t \to p(\omega_t(1))$. Consider the square I^2 mapped into E as indicated in the diagram. The horizontal segments are mapped by ω_t and the whole is a

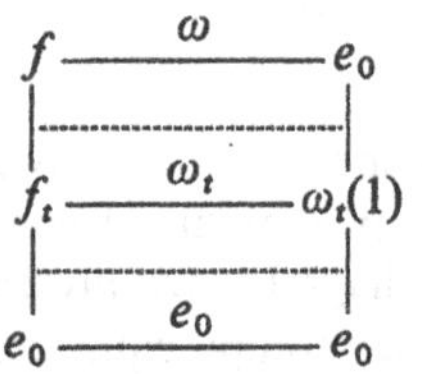

continuous map $I^2 \to E$ which essentially represents φ'. When we project this square into B by means of p, the left edge and bottom edge and the upper-right corner are mapped to the base point $b_0 = p(e_0)$. This square then yields a homotopy between $t \to p(\omega(1-t))$ and $t \to \omega_t(1)$, represented in the diagram by imagining the square being swept out by a radial line rotating about the upper-right corner. This homotopy completes the proof of the lemma.

EXERCISE

1. Let $f\colon S^3 \to K(Z,3)$ generate $H^3(S^3)$. By the construction of the fibre mapping sequence, we may assume a diagram $X \xrightarrow{p} S^3 \xrightarrow{f} K(Z,3)$, where X is the fibre of f. Converting p to a fibration, we have a fibre space (X,p,S^3) where the total space X is 3-connected and p induces isomorphisms $\pi_i(X) \approx \pi_i(S^3)$ for all $i > 3$. Use this to show that the smallest integer n such that $\pi_n(S^3)$ contains an element of order p, where p is an odd prime, is $n = 2p$. (Find the cohomology of the fibre, write the cohomology spectral sequence, and apply the universal coefficient theorems and the Hurewicz theorem mod $\mathcal{C}_p$.)

REFERENCES

The fibre mapping sequence

1. F. P. Peterson [1].

The inclusion mapping sequence

1. M. G. Barratt [1].
2. D. Puppe [1].

CHAPTER 15

PROPERTIES OF THE STABLE RANGE

From time to time we have considered various phenomena which exhibit particularly regular behavior in a certain range of dimensions depending on the connectivity of the spaces in question. Perhaps the most fundamental example is the suspension theorem (Theorem 1 of Chapter 12), which asserts that the suspension homomorphism $E: \pi_i(X) \to \pi_{i+1}(SX)$ is an isomorphism, provided $i \leq 2n-2$, where X is $(n-1)$-connected. Stated briefly, the theorem asserts that the suspension homomorphism E is an isomorphism "in the stable range."

In this chapter we examine various constructions which have a stable range and we point out exactly what nice property holds and in what range. In future chapters when we wish to restrict attention to the stable range, we merely utter the magic phrase and usually leave to the reader the task of working out exactly what dimensional hypotheses are necessary.

NATURAL GROUP STRUCTURES

We first recall the ways in which a set of homotopy classes $[K,X]$ may have a natural group structure; namely, K a suspension or X a space of loops. Speaking precisely, the pinching map $SK \to SK \vee SK$ induces a group operation on $[SK,X]$, natural with respect to maps of X, while the multiplication $\Omega X \times \Omega X \to \Omega X$ induces a group structure on $[K,\Omega X]$, natural with respect to maps of K. On the other hand, we recall also that the sets $[SK,X]$ and $[K,\Omega X]$ are naturally equivalent; this equivalence is a group isomorphism as well. This observation makes trivial the proof of the following result.

Proposition 1

Let $f: L \to K$ and $g: X \to Y$. Then the induced maps $(Sf)^{\#}: [SK,X] \to [SL,X]$ and $(\Omega g)_{\#}[K,\Omega X] \to [K,\Omega Y]$ are group homomorphisms.

For our mapping sets to be abelian groups, we consider mapping sets of the form $[S^2K,X]$, $[SK,\Omega X]$, and $[K,\Omega^2 X]$. We ask the reader to verify that each has an abelian group structure, that the three are naturally isomorphic as groups, and that maps $SL \to SK$ or $\Omega X \to \Omega Y$ induce group homomorphisms (see Exercise 1).

Now suppose X is $(n-1)$-connected. The identity map of $S^2 X$ induces a map $h: X \to \Omega^2 S^2 X$; by the suspension theorem, h induces an isomorphism in homotopy through dimension $2n-2$. Thus by Theorems 2 and 3 of Chapter 13, dim $K \leq 2n-2$ implies that $h_{\#}: [K,X] \to [K,\Omega^2 S^2 X]$ is a one-to-one correspondence. Hence by transport of structure we have the following.

Proposition 2

Suppose X is $(n-1)$-connected and dim $K \leq 2n-2$. Then $[K,X]$ has an abelian group structure, natural with respect to maps $f: L \to K$ and $g: X \to Y$, where Y is $(n-1)$-connected and dim $L \leq 2n-2$.

Briefly stated, in the stable range $[K,X]$ has a natural abelian group structure.

EXAMPLES: COHOMOLOGY AND HOMOTOPY GROUPS

Recall by Theorem 1 of Chapter 1 that we have a one-to-one correspondence between $[X,K(\pi,n)]$ and $H^n(X;\pi)$. Viewing $K(\pi,n)$ as $\Omega K(\pi, n+1)$, we obtain a group structure on $[X,K(\pi,n)]$, natural with respect to maps of X. The correspondence is indeed an isomorphism of groups—in fact, $[X,K(\pi,n)]$ is probably the most suitable *definition* of $H^n(X;\pi)$.

Now let $\theta \in H^q(\pi,n; G)$ be a cohomology operation of type $(\pi,n; G,q)$. Is $\theta: H^n(X;\pi) \to H^q(X;G)$ necessarily a group homomorphism? In general, the answer is no. (Consider, for example, the operation taking u to u^2 on two-dimensional classes in the integral cohomology ring of $S^2 \times S^2$.) However, Proposition 1 tells us that the answer is yes, provided θ can be written in the form $\Omega\psi$ for some $\psi: K(\pi, n+1) \to K(G, q+1)$, i.e., provided θ is in the image of the cohomology suspension.

We say that θ is *additive* if $\theta\colon H^n(X;\pi) \to H^q(X;G)$ is a group homomorphism for every X. Summarizing the above remarks, we state the following.

Proposition 3

Suppose θ is in the image of the cohomology suspension. Then θ is additive.

We next consider the import of Proposition 1 for homotopy. Since $\pi_i(X)$ is just $[S^i,X]$, the group structure in π_i arises from viewing S^i as a suspension for $i \geq 1$; π_i is abelian for $i \geq 2$ since S^i is then a double suspension. The naturality with respect to maps $g\colon X \to Y$ is just the statement that $g_\#\colon \pi_i(X) \to \pi_i(Y)$ is a homomorphism.

Now let $f\colon S^p \to S^i$. Then composition with f induces a natural map, also denoted f, from $\pi_i(X)$ to $\pi_p(X)$. By analogy this transformation is called a *primary homotopy operation.* Need such a primary homotopy operation be additive? In general, the answer is no, but Proposition 1 tells us that the answer is yes if f is a suspension element.

As a very important special case, take X also to be a sphere, say, $X = S^q$. Suppose $p \leq 2i - 1$, so that f is perforce a suspension. Then the composite $g \circ f$ (where $g\colon S^i \to S^q$) is a bilinear function of the variables and thus defines a pairing $\pi_i(S^q) \otimes \pi_p(S^i) \to \pi_p(S^q)$. This pairing is clearly compatible with suspension, that is, $E(g \circ f) = E(g) \circ E(f)$. Therefore the pairing is well-defined on the stable stems.

We denote the stable k-stem by G_k. Thus $G_k = \pi_{n+k}(S^n)$ for n sufficiently large. Let G be the graded group with G_k as its kth term. We now may state the following theorem.

Theorem 1

The composition operation endows G with the structure of commutative graded ring with unit.

By the above remarks, and by the observation that the identity map of the n-sphere serves as unit in G_0, only the commutativity need be proved. We omit the proof of this result of Barratt and Hilton but remind the reader that commutativity for a graded ring involves a sign.

In Chapter 17 we will compute the product structure in the 2-component of G in the range of the calculations of Chapter 12, namely, through degree 7.

CONSEQUENCES OF SERRE'S EXACT SEQUENCE

We now recall Theorem 2 of Chapter 8 and its analogue for cohomology. Suppose $p\colon E \to B$ is a fibration with fibre F. If B is $(p-1)$-connected and F is $(q-1)$-connected, then we have the exact sequence

$$H_{p+q-1}(F) \xrightarrow{i_*} H_{p+q-1}(E) \xrightarrow{p_*} H_{p+q-1}(B) \xrightarrow{\tau} H_{p+q-2}(F) \rightarrow \cdots$$

and similarly in cohomology.

Speaking roughly, we would say that Serre's exact sequence is valid in the stable range.

We now apply Serre's exact sequence to the cohomology suspension, which we define for an arbitrary space X as follows. Let $u \in H^{n+1}(X;\pi)$ be represented by the map $u: X \rightarrow K(\pi, n+1)$. Then the cohomology suspension of u, ${}^1u \in H^n(\Omega X;\pi)$, is the class represented by the map Ωu.

To study the cohomology suspension, consider the contractible fibration over X with fibre ΩX. In the range of Serre's exact sequence, the suspension and the transgression are inverse homomorphisms (see Exercise 2). We will obtain the following.

Proposition 4

Suppose X is m-connected. Then the cohomology suspension

$${}^1(\ ): H^{k+1}(X) \rightarrow H^k(\Omega X)$$

is an isomorphism if $k \leq 2m-1$.

PROOF: We apply Serre's exact sequence with $B = X$, $F = \Omega X$, and E contractible. Then $p = m-1$ and $q = m$. The isomorphism holds for $k \leq p - q - 2 = 2m - 1$.

Corollary 1

Let $\theta \in H^k(\pi,m; G)$. Suppose $k \leq 2m-1$. Then we may write $\theta = {}^1\psi$ in a unique way. Hence θ is additive.

Briefly put, in the stable range a cohomology operation is additive and may be uniquely desuspended.

For $u \in H^k(Y;\pi)$, we denote by $E(u)$ the total space of the fibration induced over Y by u from the contractible fibration over $K(\pi,k)$.

Proposition 5

Suppose X is m-connected and $n \leq 2m-1$. Let $u \in H^n(\Omega X;\pi)$. Then $E(u)$ has the homotopy type of a loop space.

PROOF: By Proposition 4, we may write $u = {}^1v$ for some $v \in H^{n+1}(X;\pi)$. Then $E(u)$ and $\Omega E(v)$ have the same homotopy type.

Moreover, by the results of the Appendix to Chapter 14 together with Proposition 1, we note that for each K, the map $[K,E(u)] \rightarrow [K,\Omega X]$ is a homomorphism.

Corollary 2

Suppose X is $(n-1)$-connected and $\pi_i(X)=0$ for $i \geq 2n-1$. Then X has the homotopy type of a space of loops.

PROOF: Consider a Postnikov system for X with typical term X_j. We work by induction on j. Now X_{j+1} is of the form $E(u)$ for a certain class $u \in H^p(X_j;\pi)$ for some π, where $p \leq 2n-1$. By the induction hypothesis, X_j is a space of loops (on an n-connected space, since X is $(n-1)$-connected). Therefore Proposition 5 applies and $E(u) = X_{j+1}$ is a loop space. But $X = X_{2n}$, and thus the corollary is proved.

We next consider the relation between the fibre and cofibre of a map. Suppose $p: E \rightarrow B$ is a mapping. By the standard construction, we may assume p is a fibre map with fibre F; by another standard construction, we may assume p is an inclusion with "cofibre" B/E.

There is a natural map $f: SF \rightarrow B/E$, induced from the following diagram, since SF is the cofibre of $F \rightarrow *$.

$$\begin{array}{ccc} F & \rightarrow & E \\ \downarrow & & \downarrow \\ * & \rightarrow & B \end{array}$$

Theorem 2

Suppose B is $(p-1)$-connected and F is $(q-1)$-connected. Then the natural map f induces an isomorphism from $\pi_i(SF)$ to $\pi_i(B/E)$ for $i \leq p+q-2$.

PROOF: We implicitly assume that p and q are sufficiently large to assure that SF and B/E are simply connected. It is thus enough to prove that f induces an isomorphism in homology for $i \leq p+q-1$. To do so, we merely compare the Serre exact homology sequence for the fibration, replacing F by SF with a dimension shift, with the exact homology sequence of the pair (B,E). With the help of f, we then have a homomorphism of long exact sequences, with the terms in question sandwiched between the common terms for E and B. The result follows from the five-lemma, noting that the last term to be sandwiched is $H_{p+q-2}(F) = H_{p+q-1}(SF)$.

Briefly put, in the stable range the cofibre and fibre agree, except for a dimension shift.

We are already familiar with the following basic facts (see Chapter 14): cofibre commutes with suspension and fibre commutes with loops. Indeed, these are the properties which give rise to the cofibre (inclusion) and fibre mapping sequences, respectively. But Theorem 2 tells us that fibre and cofibre agree (except for the dimension shift) in the stable range. Thus in the stable range cofibre commutes with loops and fibre commutes with suspension.

We leave to the reader the task of precise formulation of the numerology for these statements. Nonetheless we will use them, especially in Chapter 18. For example, we feel free to use the fibre mapping sequence of the pair (B,E) of form

$$\cdots \rightarrow [K,\Omega^r E] \rightarrow [K,\Omega^r B] \rightarrow [K,\Omega^r(B/E)] \rightarrow \cdots$$

which is valid when r and dim K are sufficiently small.

DISCUSSION

What we have done in this chapter is to establish a stable range in which many things are well-behaved. In Chapter 16 we will study a new sort of cohomology operation in the stable range; in Chapter 17 we will apply these operations to gain new information about the composition ring G of stable homotopy groups of spheres. Stability becomes even more crucial in Chapter 18, where we introduce a totally different tool for the study of stable mapping groups.

We based our comparison of fibre and cofibre on the natural map $f\colon SF \rightarrow B/E$. The adjoint of f is another natural map $g\colon F \rightarrow \Omega(B/E)$. The study of g would be a dual method of approach to the subject. Our proof of Theorem 2 uses in a key step the exact homology sequence of the fibration in the stable range. A completely dual proof of the corresponding results would require an exact homotopy sequence of the cofibration $E \rightarrow B \rightarrow B/E$. Thus we would need, in the stable range, an excision property for homotopy in order to identify $\pi_i(B,E)$ with $\pi_i(B/E)$. That such an excision property holds in the stable range is a consequence of the triad theorem of Blakers and Massey.

It is possible to formulate precisely a "stable category" in which stability conditions are built in. The objects in this category are known as *spectra*; they were first introduced by Lima. Subsequently Spanier, G. W. Whitehead, and Kan have put stable homotopy theory, via spectra, on a firm footing.

The use of spectra leads, as one might well imagine, to enormous simplifications in the formalism; no longer need every statement be qualified by an awkward range of dimensions which have to be kept track of. On the other hand, there is a great deal of technique needed to establish the theory.

For the applications we give, the mass of technique outweighs the gain; thus we do not introduce spectra here. However, for more sophisticated applications they seem to be indispensable.

EXERCISES

1. The mapping sets $[S^2K,X]$, $[SK,\Omega X]$, and $[K,\Omega^2 X]$ are in natural one-to-one correspondence. The first and third have group structures; the second has two. Prove that these group structures all coincide under the natural set isomorphisms and are abelian. Prove also that maps $SL \to SK$ and $\Omega X \to \Omega Y$ induce group homomorphisms.
2. In the range of Serre's exact sequence, the groups $H^{k+1}(X)$ and $H^k(\Omega X)$ are isomorphic. Prove that the cohomology suspension and cohomology transgression are indeed inverse isomorphisms. HINT: Recall that the coboundary operator for a pair may be defined by mapping the inclusion mapping sequence (Chapter 14) of the pair into an appropriate Eilenberg-MacLane space and use the second definition of transgression (Chapter 8).

REFERENCES

Spectra

1. J. F. Adams [4].
2. D. M. Kan [1,2].
3. — and G. W. Whitehead [1].
4. G. W. Whitehead [5].

Commutativity of the composition ring G

1. M. G. Barratt and P. J. Hilton [1].

The triad theorem

1. A. L. Blakers and W. S. Massey [1].

CHAPTER 16

HIGHER COHOMOLOGY OPERATIONS

We are moving toward a more intimate knowledge of the homotopy groups of spheres. Before launching into the homotopy theory, we need to develop some more sophisticated tools. In this chapter we will describe two kinds of higher cohomology operations which will be useful in the homotopy calculations of the following chapter.

To simplify matters we will restrict our attention to the "stable range." In each specific context the precise limits of the "stable range" will be more or less evident from the context; sometimes we will make them explicit.

FUNCTIONAL COHOMOLOGY OPERATIONS

The first new kind of cohomology operation is the *functional operation.* We suppose that we are given a map $f: Y \to X$ (this is the "function") and a primary operation θ of type $(G,n;\pi,q)$. For stability, we assume $q \leq 2n-2$, so that θ is in the image of the cohomology suspension and thus θ is a homomorphism when considered as a transformation of additive groups $H^n(\ ;G) \to H^q(\ ;\pi)$.

Since every map is homotopically equivalent to an inclusion map, we may as well assume that f is an inclusion map and $Y \subset X$.

Let u be a cohomology class in $H^n(X;G)$ satisfying the two conditions $f^*u = 0 \in H^n(Y;G)$ and $\theta(u) = 0 \in H^q(X;\pi)$. Consider the following diagram, in which each row is exact.

Diagram 1

$$\begin{array}{ccccccc} & & H^{n-1}(Y;G) & \xrightarrow{\delta} & H^n(X,Y;G) & \xrightarrow{j^*} & H^n(X;G) & \xrightarrow{f^*} & H^n(Y;G) \\ & & \downarrow{}^1\theta & & \downarrow\theta & & \downarrow\theta & & \\ H^{q-1}(X;\pi) & \xrightarrow{f^*} & H^{q-1}(Y;\pi) & \xrightarrow{\delta} & H^q(X,Y;\pi) & \xrightarrow{j^*} & H^q(X;\pi) & & \end{array}$$

Since $f^*u=0$, we can find $u' \in H^n(X,Y;G)$ such that $j^*(u')=u$. Since $\theta(u)=0$, $j^*\theta(u')=\theta j^*(u')=\theta(u)=0$, and so we can find $u'' \in H^{q-1}(Y;\pi)$ such that $\delta(u'')=\theta(u')$. Of course u'' is not uniquely determined by u; what is uniquely determined by u is the projection of u'' in $H^{q-1}(Y;\pi)/Q$, where the *indeterminacy* Q is the sum ${}^1\theta(H^{n-1}(Y;G))+f^*(H^{q-1}(X;\pi))$ of subgroups of $H^{q-1}(Y;\pi)$. To see this, suppose that v' and v'' were chosen in place of u' and u''. Then $j^*(u'-v')=0$, and so $u'-v'=\delta b$ for some $b \in H^{n-1}(Y;G)$. Therefore $\delta(u''-v'')=\theta(u')-\theta(v')=\theta(\delta b)=\delta({}^1\theta(b))$, so that $(u''-v''-{}^1\theta(b))$ is in the kernel of δ and thus $u''-v''=f^*(a)+{}^1\theta(b)$ for some $a \in H^{q-1}(X;\pi)$ and some $b \in H^{n-1}(Y;G)$.

We define the functional operation θ_f by putting $\theta_f(u)$ to be the coset $u''+Q$. Thus θ_f has as its domain a subgroup of $H^n(X;G)$, namely, the intersection of $\ker f^*$ and $\ker \theta$; and it takes values in the quotient group $H^{q-1}(Y;\pi)/Q$, where $Q=\operatorname{im}{}^1\theta+\operatorname{im} f^*$. Like the operation θ, it goes from cohomology with coefficients in G to cohomology with coefficients in π; note, however, that it raises dimension by only $q-n-1$ and not, like θ, by $q-n$. Like the homomorphism f^*, θ_f goes from the cohomology of X to that of Y.

These functional operations are natural in the following sense. Suppose we have the following commutative diagram and suppose that $u \in H^n(X;G)$

$$\begin{array}{ccc} Y & \xrightarrow{f} & X \\ \uparrow\eta & & \uparrow\xi \\ Y' & \xrightarrow{f'} & X' \end{array}$$

satisfies $f^*u=0$, $\theta u=0$, so that $\theta_f(u)$ is defined in $H^{q-1}(Y;\pi)/Q$. Then obviously $(f')^*(\xi^*(u))=0$ and $\theta(\xi^*(u))=0$, so that $\theta_{f'}$ is defined on $\xi^*(u)$, taking its value in $H^{q-1}(Y';\pi)/Q'$. Then we claim that $\eta^*(\theta_f(u))= \theta_{f'}(\xi^*(u))$ or, more precisely, that the left-hand member of this "equation" is contained in the right-hand member when both are considered as subsets (cosets) of $H^{q-1}(Y';\pi)$. In fact, the indeterminacy of $\eta^*(\theta_f(u))$ is clearly

$$\eta^*(Q)=\eta^*f^*(H^{q-1}(X;\pi))+\eta^*({}^1\theta(H^{n-1}(Y;G)))$$

while that of $\theta_{f'}(\xi^*(u))$ is

$$Q'=(f')^*(H^{q-1}(X';\pi))+{}^1\theta(H^{n-1}(Y';G))$$

so that clearly $\eta^*(Q)$ is contained in Q'. To verify that η^* maps any representative of $\theta_f(u)$ to a representative of $\eta_{f'}(\xi^*(u))$ is no more than an exercise in diagram-chasing, which we leave to the reader.

One of the important applications of functional operations is in establishing that a given map is essential. It is a classical pattern of proof in algebraic topology to show that a map is homotopically non-trivial, i.e., essential, by showing that it is algebraically non-trivial. Thus if a map $f: Y \to X$ induces a non-zero homomorphism $f^*: H^*(X) \to H^*(Y)$, then clearly f must be essential. The following proposition represents a considerable sharpening of this method.

Proposition 1

Let f be a map $Y \to X$. Suppose there exists a primary operation θ and a cohomology class u such that $\theta_f(u)$ is defined and non-zero. Then f is essential.

Of course to say that $\theta_f(u)$ is non-zero means to say that it is not the zero coset.

Under the conditions of the proposition, we say that the operation θ *detects* the map f.

To prove the proposition, suppose that f is null-homotopic. We may as well assume that f is an inclusion map. Then the fact that $f \simeq 0$ means that $X \cup_f CY$ has the homotopy type of the wedge $X \vee SY$, and the cohomology sequence of the pair (X, Y) splits: $H^q(X, Y; \pi) \approx H^q(X;\pi) \oplus H^{q-1}(Y;\pi)$. The splitting map r^* gives a commutative diagram

$$\begin{array}{ccc} H^n(X, Y; G) & \xleftarrow{r^*} & H^n(X;G) \\ \downarrow{\scriptstyle\theta} & & \downarrow{\scriptstyle\theta} \\ H^q(X, Y; \pi) & \xleftarrow{r^*} & H^q(X;\pi) \end{array}$$

where θ is an *arbitrary* primary operation of type $(G,n; \pi,q)$. Then, if $u \in H^n(X;G)$ satisfies $f^*(u) = 0$ and $\theta(u) = 0$, we can represent $\theta_f(u)$ by zero, since we can take $u' = r^*(u)$ and then $\theta(u') = \theta(r^*(u)) = r^*(\theta(u)) = 0$. This proves the proposition.

As an example, let f be the map $S^{n+1} \to S^n$ which is the $(n-2)$-fold suspension of the Hopf map $\eta: S^3 \to S^2$, and let θ be the operation Sq^2. Recall (from Chapter 4) that Sq^2 is non-zero in the complex $K = S^n \cup_f e^{n+2}$. Now K is precisely M/S^{n+1}, where M is the mapping cylinder of f. Then M has the homotopy type of S^n. We take $Y = S^{n+1}$, $X = M \simeq S^n$, and of course $G = \pi = Z_2$. Let u be the generator of $H^n(X;Z_2)$. Obviously $Sq^2u = 0$ and $f^*u = 0$. Therefore $Sq^2_f(u)$ is defined. The diagram defining it is as shown below. (All coefficients are understood to be in Z_2). We have

$$\begin{array}{ccccccc} & & H^{n-1}(Y) & \longrightarrow & H^n(K) & \xrightarrow{j} & H^n(X) \rightarrow H^n(Y) \\ & & \downarrow & & \downarrow Sq^2 & & \downarrow \\ H^{n+1}(X) & \rightarrow & H^{n+1}(Y) & \xrightarrow{\delta} & H^{n+2}(K) & \rightarrow & H^{n+2}(X) \end{array}$$

$H^n(K) \approx H^n(X) = Z_2$, since $H^k(Y) = 0$ for $k \neq n+1$; thus we must take u' to be the generator of this group. Then Sq^2u' is non-zero. Finally, since $H^k(X) = 0$ for $k \neq n$, the δ shown is an isomorphism and we must take u'' to be the generator of $H^{n+1}(Y)$. The indeterminacy is zero, and we have proved that $Sq^2_f(u)$ is non-zero. Thus, in the language introduced above, Sq^2 detects f (as an essential map). Note that f^* is zero in all dimensions.

In the same way, the other Hopf maps ν: $S^{n+3} \to S^n$ and $\sigma: S^{n+7} \to S^n$ give non-trivial functional operations Sq^4_ν and Sq^8_σ, respectively.

ANOTHER FORMULATION OF θ_f

We now introduce another (equivalent) definition of the functional operation θ_f. Let θ, f, and u be as before. We can use θ to define a fibre space E over $K(G,n)$ induced from the standard contractible fibre space over $K(\pi,q)$, as indicated in the following diagram.

Diagram 2

$$\begin{array}{ccccccc} F = K(\pi, q-1) & & \xrightarrow{i} & & E & & \\ \uparrow u'' & & & \overset{\tilde{u}}{\nearrow} & \downarrow p & & \\ Y & \xrightarrow{f} & X & \xrightarrow{u} & B = K(G,n) & \xrightarrow{\theta} & K(\pi,q) \end{array}$$

To say that $\theta(u) = 0$ is to say that the map u can be lifted to give a map $\tilde{u}$: $X \to E$ such that $p\tilde{u} = u$. To say that $f^*(u) = 0$ is to say that the composite uf is null-homotopic. This means that $\tilde{u}f$: $Y \to E$ is in the kernel of $p_\#$: $[Y,E] \to [Y,B]$, which is the same as the image of $i_\#$: $[Y,F] \to [Y,E]$, so that there exists a map u'': $Y \to F$ such that iu'' is homotopic to $\tilde{u}f$. The corresponding cohomology class $u'' \in H^{q-1}(Y;\pi)$ will represent the image of u under the functional operation θ_f, which we shall denote $\bar{\theta}_f$ for the time being to distinguish it from the θ_f previously defined. It is easiest to evaluate the indeterminacy in the definition of $\bar{\theta}_f$ by considering the following diagram.

Diagram 3

$$\begin{array}{ccccccc} & & H^{q-1}(X;\pi) & \rightarrow & [X,E] & \rightarrow & H^n(X;G) \xrightarrow{\theta} H^q(X;\pi) \\ & & \downarrow & & \downarrow & & \downarrow \\ H^{n-1}(Y;G) & \xrightarrow{1\theta} & H^{q-1}(Y;\pi) & \rightarrow & [Y,E] & \rightarrow & H^n(Y;G) \end{array}$$

Here the vertical maps are f^* and each row is exact (cf. Chapter 14). Since we are in the stable range, Diagram 3 is a diagram of groups and homomorphisms. Now the definition of $\bar{\theta}_f(u)$ which was given above is easily traced on this diagram; formally, it bears the closest possible resemblance to the definition of $\theta_f(u)$. In particular, it is evident, chasing the diagram, that the indeterminacy is $f^*(H^{q-1}(X;\pi)) + {}^1\theta(H^{n-1}(Y;G))$, which is exactly the same as the indeterminacy Q in the definition of θ_f. Thus $\bar{\theta}_f(u)$ may be considered as a well-defined element of the quotient group $H^{q-1}(Y;\pi)/Q$.

We leave it to the reader to verify that such operations $\bar{\theta}_f$ are natural in the same sense as we proved before, using the first definition of functional operations. This fact would be immediate from the next proposition, but we want to use it in the proof, so it must be established directly.

Proposition 2

The functional operations θ_f and $\bar{\theta}_f$ are identical.

We will prove this by the method of universal example. For the universal example, take $X = E$ (where E is the total space introduced above), $Y = F = K(\pi, q-1)$, $f = i: F \to E$, and $u = p \in H^n(E;G)$. Then in Diagram 2, we may take u and u'' to be the appropriate identity maps, and thus it is obvious that the identity map (or fundamental class) $\iota_{q-1} \in H^{q-1}(Y;\pi)$ represents $\bar{\theta}_f(u)$. To see that ι_{q-1} also represents $\theta_f(u)$, consider the following portion of Diagram 1:

$$\begin{array}{ccc} & H^n(E,F;G) & \xrightarrow{j^*} H^n(E;G) \\ & \downarrow{\scriptstyle\theta} & \\ H^{q-1}(F;\pi) \xrightarrow{\delta} & H^q(E,F;\pi) & \end{array}$$

Now in the stable range the cohomology exact sequence of Serre for this fibre space over $B = K(G,n)$ yields isomorphisms

$$H^k(E,F;\ ) \approx H^k(B;\ )$$

for whatever coefficients. Under this isomorphism, the class $p \in H^n(E,F;G)$ corresponds to the fundamental class $\iota_n \in H^n(B;G)$ and the coboundary $\delta: H^{q-1}(F;\pi) \to H^q(E,F;\pi)$ corresponds to the transgression $\tau: H^{q-1}(F;\pi) \to H^q(B;\pi)$. If we choose $p' \in H^n(E,F;G)$ such that $j^*(p') = p$, then $\theta(p')$ corresponds to $\theta(\iota_n)$. Thus, showing that ι_{q-1} represents $\theta_f(u)$ comes down to showing that $\tau(\iota_{q-1}) = \theta(\iota_n)$. But by Formula 1 of Chapter 11 this is immediate from the construction of E. Thus the proposition is established for the universal example.

The general case follows easily by naturality, as in the following calculation:

$$\begin{aligned} \theta_f(u) = \theta_f(\tilde{u}^*(p)) &= (\bar{\theta}_f u)^*(\theta_i(p)) && \text{by naturality of } \theta_f \\ &= (\bar{\theta}_f u)^*(\bar{\theta}_i(p)) && \text{by the universal example} \\ &= \bar{\theta}_f(\tilde{u}^*(p)) && \text{by naturality of } \bar{\theta}_f \\ &= \bar{\theta}_f(u) \end{aligned}$$

This completes the proof of Proposition 2.

SECONDARY COHOMOLOGY OPERATIONS

The other new kind of cohomology operation that we will be using is the *secondary operation*. We suppose that we are given a two-stage Postnikov system as in the following diagram.

Diagram 4

$$\begin{array}{ccccc} F = K(\pi, q-1) & \xrightarrow{i} & E & \xrightarrow{\varphi} & K = K(H,m) \\ & \nearrow^{\tilde{u}} & \downarrow p & & \\ X & \xrightarrow{u} & B = K(G,n) & \xrightarrow{\theta} & K(\pi,q) \end{array}$$

Here θ is a given primary operation of type $(G,n;\pi,q)$. Suppose also we are given a cohomology class φ in $H^m(E;H)$. The secondary operation Φ will be defined on a class $u \in H^n(X;G)$ provided that the composition θu is null-homotopic, i.e., that $\theta(u) = 0$. In this event we can find a lifting $\tilde{u}$: $X \to E$ such that $p\tilde{u} = u$. We can therefore consider the class $\tilde{u}^*(\varphi) \in H^m(X;H)$. The indeterminacy is due to the choice of $\tilde{u}$; the kernel of $p_\#: [X,E] \to [X,B]$ is the image of $i_\#$ $[X,F] \to [X,E]$, and so it is easy to see that $\tilde{u}^*(\varphi)$ may vary by the addition of an arbitrary element of the image of $i^*\varphi$: $H^{q-1}(X;\pi) \to H^m(X;H)$. Thus the coset $\tilde{u}^*(\varphi) + \text{im}\,(i^*\varphi)$ is well-defined. We thereby obtain a secondary operation Φ which has for its domain $\ker \theta \subset H^n(X;G)$ and which takes values in coker $(i^*\varphi)$, a quotient group of $H^m(X;H)$.

We have made use of the stability assumption in the assertion that the indeterminacy is the subgroup im $(i^*\varphi)$. What is needed is that q and m are not too large with respect to n or, precisely, that q and m are each less than $2n - 1$. Without such an assumption, the evaluation of the indeterminacy becomes a thorny problem.

We remark that if φ is such that $i^*\varphi = 0$, or equivalently if φ is in the image of $p^*: H^m(B;H) \to H^m(E;H)$, then Φ is essentially a primary operation, in the following sense. Let $\varphi = p^*\theta'$ where $\theta' \in H^m(B;H)$; then we have a commutative diagram of maps $X \to K = K(H,m)$ with $\varphi\tilde{u} = \theta' u$, and consequently $\Phi(u) = \tilde{u}^*(\varphi) = u^*(\theta') = \theta'(u)$, with zero indeterminacy, and so Φ is nothing but the restriction of θ' to ker θ.

We will therefore assume from now on that $i^*\varphi$ is non-zero.

There is an obvious naturality formula for secondary operations: if f is a map $Y \to X$, then $f^*(\Phi(u)) = \Phi(f^*(u))$, modulo the indeterminacy of the right-hand side. To see this, consider Diagram 4 augmented by the map $f: Y \to X$. The class of the map $\varphi\tilde{u}$ represents $\Phi(u) \in H^m(X;H)/Q$, where $Q = (\varphi i)[X,F] = \text{im } (i^*\varphi)$. Then $f^*(\Phi(u))$ is represented by $\varphi\tilde{u}f$ and has indeterminacy $Q' = f^*Q = f^* \text{ im } (i^*\varphi)$. We could equally well represent Q' as $(\varphi i)(\text{im} f^*)$, where $f^*: H^m(X;H) \to H^m(Y;H)$ or equivalently $f^* = f^\#: [X,F] \to [Y,F]$. Now $\tilde{u}f$ is a lifting of uf, and thus obviously the class $\varphi\tilde{u}f$ also represents $\Phi(f^*(u))$. The indeterminacy of this term, however, is $Q'' = (\varphi i)[Y,F] = \text{im}\,(i^*\varphi)$, and clearly $Q' \subset Q''$. Thus the naturality formula must be interpreted as an equality in $H^m(X;H)/Q''$.

SECONDARY OPERATIONS AND RELATIONS

In a very precise sense, a secondary operation corresponds to a relation between primary operations. Suppose first that a secondary operation Φ is given, as above. Under suitable stability hypotheses, the composition $(\varphi i): K(\pi, q-1) \to K(H,m)$ has a unique representation as ${}^1\psi$ where $\psi \in H^{m+1}(\pi,q;H)$. We therefore have the following diagram.

Diagram 5

$$\begin{array}{ccccc} F = K(\pi, q-1) & \xrightarrow{i} & E & \xrightarrow{\varphi} & K(H,m) \\ & & \downarrow{\scriptstyle p} & & \\ & B = K(G,n) & \xrightarrow{\theta} & K(\pi,q) & \xrightarrow{\psi} K(H, m+1) \end{array}$$

where $(\varphi i) = {}^1\psi$. The diagram

$$\begin{array}{ccc} H^{q-1}(F;\pi) & \xrightarrow{\tau} & H^q(B;\pi) \\ \downarrow{\scriptstyle {}^1\psi} & & \downarrow{\scriptstyle \psi} \\ H^m(F;H) & \xrightarrow{\tau} & H^{m+1}(B;H) \end{array}$$

is commutative (under suitable sign conventions); we omit the proof of this general lemma. Granting this, we have

$$0=\tau(i^*(\varphi))=\tau(\varphi i(\iota_{q-1}))=\tau({}^1\psi(\iota_{q-1}))=\psi\tau(\iota_{q-1})=\psi\theta$$

and we thus obtain the relation $\psi\theta=0$ between the primary operations θ and ψ.

Conversely, suppose we begin with a relation $\psi\theta=0$. Using θ, we construct the fibre space $F\to E\to B$ as before. Since $\tau({}^1\psi(\iota_{q-1}))=\psi\theta=0$, ${}^1\psi$ must be in the image of i^*, so that we have ${}^1\psi=i^*(\varphi)$ for some $\varphi\in H^m(E;H)$. This completes the reconstruction of the original diagram, and we can define a secondary operation Φ. Note, however, that φ is not unique; we may vary φ by adding an arbitrary element in the image of p^*. Thus the relation $\psi\theta=0$ does not determine Φ uniquely; but any two Φ's thus obtained will differ only by a primary operation, according to the previous remark.

Observe the relationship of degrees:

$$\psi\theta\colon H^n(X;G)\to H^{m+1}(X;H)$$

whereas Φ goes from a subgroup of $H^n(X;G)$ to a quotient group of $H^m(X;H)$.

Before giving an example of a secondary operation, we remark that the theory can readily be generalized by replacing the space $K(\pi,q)$ by a cartesian product of such spaces, $\prod_i K(\pi_i,q_i)$. Then the loop space, F, is similarly a product. The operation θ becomes, in the general formulation, an ordered n-tuple $(\theta_1,\ldots,\theta_n)$, since a map of B into $\prod_i K_i$ (where $K_i=K(\pi_i,q_i)$) is just an n-tuple of maps $\theta_i\colon B\to K_i$.

Our example will be of this kind. Consider the Adem relation $R\colon Sq^3Sq^1+Sq^2Sq^2=0$. From this relation we can obtain a secondary operation, as follows. We take the diagram

$$\begin{array}{l} F=K(Z_2,n)\times K(Z_2,n+1)\to E \\ \qquad\qquad\qquad\qquad\qquad\qquad\downarrow \\ \qquad\qquad\qquad\qquad\qquad K(Z_2,n)\xrightarrow{(Sq^1,Sq^2)}K(Z_2,n+1)\times K(Z_2,n+2)\end{array}$$

The role of ${}^1\psi$ is played by the pair of operations (Sq^3,Sq^2). The cohomology class $(Sq^3\iota_n)\otimes 1+1\otimes(Sq^2\iota_{n+1})\in H^{n+3}(F)$ (coefficients in Z_2) transgresses to zero, where the symbol 1 denotes the unit in the appropriate cohomology ring. Therefore this class is $i^*\varphi$ for some $\varphi\in H^{n+3}(E)$. In this example, φ is uniquely determined, for it is a routine matter to verify that p^* is zero on $H^{n+3}(B)$; in fact $H^{n+3}(B)$ is generated by $Sq^2Sq^1\iota_B=\tau(Sq^2\iota_n)$ and by $Sq^3\iota_B=\tau(Sq^1\iota_{n+1})$. Then if $u\in H^n(X;Z_2)$ is a class such that Sq^1u and Sq^2u are both zero, $\Phi(u)$ is a well-defined element of the quotient group $H^{n+3}(X;Z_2)/Q$ where Q is the image of the map $[X,F] \rightarrow H^{n+3}(X;Z_2)$ induced by $Sq^3 + Sq^2 : F \rightarrow K(Z_2, n+3)$. This operation Φ, discovered by Adem using entirely different methods, was one of the first secondary cohomology operations to be studied; it can be used to show that the composition $S^{n+2} \rightarrow S^{n+1} \rightarrow S^n$ is essential, where both maps are the suspensions of the Hopf map $\eta: S^3 \rightarrow S^2$.

THE PETERSON-STEIN FORMULAS

There reader has no doubt observed some similarity between the defining diagrams for $\bar{\theta}_f$ and Φ. These two sorts of operations are indeed closely related; the following formulas tell how.

Throughout this section we retain the notations and assumptions of Diagrams 4 and 5; that is, we have our cohomology classes θ, ψ, and φ satisfying ${}^1\psi = i^*\varphi$ and F, E, B, i, and p are as in the diagrams. Φ is the secondary operation determined by φ. We consider a map f: $Y \rightarrow X$ and a cohomology class $u \in H^n(X;G)$.

Now suppose that $f^*u = 0$, which is to say that the composition uf is null-homotopic. Then the naturality formula shows that $f^*(\Phi(u)) = 0$ in $H^m(X;H)/Q''$, where $Q'' = \text{im } {}^1\psi$. We can sharpen this result by reducing the indeterminacy from Q'' to $Q' = f^* \text{ im } {}^1\psi$ by the following formula.

Theorem 1 (first Peterson-Stein formula)

$f^*(\Phi(u)) = {}^1\psi\theta_f(u)$ in $H^m(X;H)/Q'$.

PROOF: $\theta_f(u)$ is represented by a map u'': $Y \rightarrow F$ such that $iu'' \simeq \tilde{u}f$. The indeterminacy of $\theta_f(u)$ is $Q^* = \text{im } {}^1\theta + \text{im} f^* \subset H^m(Y;H)$. Then $(\varphi i)(\theta_f u)$ is represented by the composite $(\varphi i u'')$, and its indeterminacy is $(\varphi i)(Q^*) = \text{im } {}^1\psi {}^1\theta + \text{im } {}^1\psi f^* = f^* \text{ Im } {}^1\psi = Q'$, since $\psi\theta = 0$. This shows that $(\varphi i)(\theta_f u)$ is equal to the class $(\varphi i u'')$ (modulo Q'). But in the proof of the naturality formula we saw that $f^*(\Phi(u))$ is equal to the class $(\varphi \tilde{u} f)$ (modulo Q'). Since $iu'' \simeq \tilde{u}f$, this proves the theorem.

The following observation will be used in the last two results of this chapter.

Lemma 1

In Diagram 5, with $(\varphi i) = {}^1\psi$ and $\psi\theta = 0$ as usual, we have $i^*\varphi = i^*(\psi_p\theta)$ modulo $(i^*(\text{im } ({}^1\psi))$, that is, modulo the image of the composite

$$\begin{array}{ccccc} H^{q-1}(E;\pi) & \xrightarrow{{}^1\psi} & H^m(E;H) & \xrightarrow{i^*} & H^m(F;H) \\ \| & & \| & & \| \\ [E,F] & \xrightarrow{(\varphi i)_*} & [E,K(H,m)] & \xrightarrow{i^*} & [F,K(H,m)] \end{array}$$

PROOF: To define $\psi_p\theta$, we replace p by an inclusion map and use the following diagram, which, in the stable range, is isomorphic in a natural

$$\begin{array}{ccc} & & H^q(B,E;\pi) \xrightarrow{j^*} H^q(B;\pi) \\ & & \downarrow{\scriptstyle \psi} \\ H^m(E;H) & \xrightarrow{\delta} & H^m(B,E;H) \end{array}$$

way to the next diagram so that $\psi_p\theta$ is represented by any class $x \in H^m(E;H)$

$$\begin{array}{ccc} & & H^{q-1}(F;\pi) \xrightarrow{\tau} H^q(B;\pi) \\ & & \downarrow{\scriptstyle {}^1\psi} \\ H^m(E;H) & \xrightarrow{i^*} & H^m(F;H) \end{array}$$

such that $i^*x = {}^1\psi(y)$ with $\tau(y) = \theta$. Now E is the fibre space induced by θ, so that $\tau(\iota_{q-1}) = \theta$, and we can conclude that $i^*(\psi_p\theta) = {}^1\psi(\iota_{q-1}) = i^*\varphi$ modulo the indeterminacy. The indeterminacy is

$$i^*(p^*H^m(B;H)) + {}^1\psi H^{q-1}(E;\pi)$$

Since $i^*p^* = 0$, this reduces to the indeterminacy stated in the lemma, and the result is proved.

Theorem 2 (second Peterson-Stein formula)

Suppose that in the diagram below the composite $\theta v f$ is null-homotopic. Then

$$\Phi(f^*v) = \psi_f(\theta v) \in H^m(Y;H)/Q, \text{ where } Q = (\varphi i)_\#[Y,F] + f^\#[X,K(H,m)].$$

$$\begin{array}{ccccccc} & & F = K(\pi, q-1) & \xrightarrow{i} & E & \xrightarrow{\varphi} & K(H,m) \\ & & & \overset{w}{\dashrightarrow} & \downarrow{\scriptstyle p} & & \\ Y \xrightarrow{f} X & \xrightarrow{v} & & & B = K(G,n) & \xrightarrow{\theta} K(\pi, q) \xrightarrow{\psi} & K(H, m+1) \end{array}$$

We have used the letter v instead of u to emphasize that we do not assume here that either θv or f^*v is zero. It is still understood that $(\varphi i) = {}^1\psi$.

PROOF: Since $\theta v f \simeq 0$, $\Phi(f^*v)$ is defined and is represented by φw, where $w\colon Y \to E$ is a lifting of (vf). The indeterminacy is $(\varphi i)_\#[Y,F]$. Now consider $\psi_f(\theta v) = \psi_f(v^*\theta)$. By the naturality formula for functional operations, $\psi_f(v^*\theta) = w^*(\psi_p\theta)$, modulo the indeterminacy of $\psi_f(v^*\theta)$, which is precisely the Q of the theorem. Thus we must compare φw, that is, $w^*\varphi$, with $w^*(\psi_p\theta)$. By the preceding lemma, $i^*\varphi = i^*(\psi_p\theta)$ modulo $i^*(\text{im }({}^1\psi))$, which implies that $\varphi = \psi_p\theta$ modulo $(\text{im }({}^1\psi) + \text{im }(p^\#))$ where $p^\#\colon [B,K(H,m)] \to [E,K(H,m)]$. We shall see that when we apply w^*, this is absorbed into Q, which will complete the proof. In fact, $w^*\varphi$ and $w^*(\psi_p\theta)$ can only differ by an element of the sum

$$w^*(\text{im}\,({}^1\psi)) + w^*(\text{im}\,(p^\#))$$

The first term is the image of the composite

$$[E,F] \xrightarrow{{}^1\psi} [E,K(H,m)] \xrightarrow{w^\#} [Y,K(H,m)]$$

which is the same as the composite

$$[E,F] \xrightarrow{w^\#} [Y,F] \xrightarrow{{}^1\psi} [Y,K(H,m)]$$

and thus the first term is contained in $(\varphi i)_\#[Y,F]$, the first term of Q. The second term is the image of $w^\# p^\#$, which is the same as $f^\# v^\#$, so that this term is contained in $f^\#[X,K(H,m)]$, the second term of Q. This shows that $w^*\varphi$ and $w^*(\psi_p\theta)$, which represent $\Phi(f^*v)$ and $\psi_f(\theta v)$, respectively, are equal modulo Q, which proves the theorem.

We could summarize the above proof by the calculation

$$\Phi(f^*v) = \Phi(w^*p) = w^*(\Phi(p)) = w^*\varphi = w^*(\psi_p\theta) = \psi_f(v^*\theta) = \psi_f(\theta v)$$

by simply ignoring any problem of indeterminacy. However, the evaluation of the indeterminacy is not only the hard part of the proof but also the main content of the theorem itself.

We now prove an important result which overlaps with the "Bockstein lemma" of Chapter 11.

Theorem 3

Let $(E',r,B';F')$ be a fibre space, let θ be a primary operation of type $(G,n;\pi,q)$, and let $u \in H^n(B';G)$ and $v \in H^{q-1}(F';\pi)$ be classes such that $\tau(v) = \theta(u) \in H^q(B';\pi)$. Suppose that $\theta ur \simeq 0$ and that $\psi\theta = 0$. Then

$$j^*(\psi_r(\theta u)) = {}^1\psi(v) = j^*(\Phi(r^*u))$$

in $H^m(F';H)/Q$, where $Q = j^*(\text{im}\,({}^1\psi))$, in other words Q is the image of the composite

$$[E',F] \xrightarrow{{}^1\psi} [E',K(H,m)] \xrightarrow{j^*} [F',K(H,m)]$$

(here F, E, B, φ, i, and θ are as in Diagram 5).

PROOF: For the universal example, take $(E',r,B';F')$ to be identical with $(E,p,B;F)$ and let u and v be the identity maps of $B = K(G,n)$ and $F = K(\pi, q-1)$. Then we are required to show that $i^*(\psi_p\theta) = i^*\varphi = i^*(\Phi(p))$ in $H^m(F;H)/Q_0$ where $Q_0 = i^*(\text{im}\,({}^1\psi))$. This is precisely the assertion of Lemma 1, together with the remark that $\varphi = \Phi(p)$ modulo im $({}^1\psi)$.

To prove the general case, we map it into the universal example, as in Diagram 6.

Diagram 6

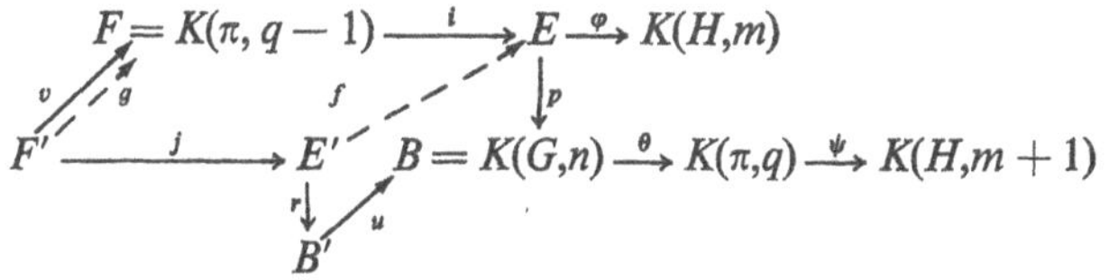

The existence of the map $f: E' \to E$ such that $pf = ur$ is implied by the fact that $\theta(ur) \simeq 0$. Then $p(fj) = u(rj)$ is trivial, so that f gives a map $g: F' \to F$ with $ig = fj$. We now have the following calculation:

$j^*(\psi_r(\theta u)) = j^*(\psi_r(u^*\theta)) = j^*f^*(\psi_p\theta)$	by the naturality formula for functional operations
$= g^*i^*(\psi_p\theta) = g^*i^*\varphi = g^*i^*(\Phi(p))$	by the result for the universal example
$= j^*f^*(\Phi(p)) = j^*\Phi(f^*p) = j^*\Phi(r^*u)$	using the naturality formula for secondary operations

Here every equality is to be interpreted as an equality in the quotient group $H^m(F';H)/Q$. We should check the indeterminacy throughout the above calculation. To begin with, $\psi_r(\theta u)$ has indeterminacy

$$Q' = {}^1\psi H^{q-1}(E';\pi) + r^*H^m(B';H)$$

and $j^*Q' = Q$. When we pass to $f^*(\psi_p\theta)$, we can only decrease the indeterminacy. When we apply the formula from the universal example, the indeterminacy is g^*Q_0; we leave it as an exercise for the reader to verify that $g^*Q_0 \subset Q$.

We have shown that $j^*(\psi_r(\theta u)) = j^*(\Phi(r^*u))$ modulo Q; it remains to compare ${}^1\psi(v)$, which is the same as φiv, whereas the middle term in the above calculation is $g^*i^*\varphi$, that is, φig. Now we do not claim that $g = v$, but we see that $\tau(g) = \tau(v)$; for, by naturality, $\tau(g) = u^*(\tau(\iota_{q-1}))$, and of course $\tau(\iota_{q-1}) = \theta(\iota_n)$ by definition of E, so that $u^*(\tau(\iota_{q-1})) = u^*\theta(\iota_n) = \theta u$, which by hypothesis equals $\tau(v)$. Thus $g = v$ modulo the image of j^*, and thus $\varphi ig = \varphi iv$ modulo Q. This completes the proof.

If, in the present theorem, we take $\theta = Sq^1 = \psi$ and $\varphi =$ the second Bockstein operator d_2, we recover the special case $r = 1$ of the Bockstein lemma of Chapter 11.

DISCUSSION

Of our two definitions of functional operations, the first, θ_f, due to Steenrod, is older and conceptually simpler. However, the formulation $\bar{\theta}_f$, due to Peterson, is much more convenient for the study of secondary operations.

The main results of this chapter, Theorems 1, 2, and 3, were proved by Peterson and Stein. The usefulness of these theorems, as we shall see in the sequel, lies in the reduction of the calculation of secondary operations to that of primary and functional primary operations.

We have given a constructive definition of secondary operations. Adams has formulated an axiomatic approach; Theorem 1 becomes his key axiom.

As the reader undoubtedly suspects, there exists a full-blown theory of higher-order operations, based at least intuitively on cohomology classes of higher-stage Postnikov systems. From the constructive viewpoint, such operations were formalized by Peterson; Maunder has successfully generalized the Adams axiomatic scheme. Such higher-order operations have been crucial in many recent calculations, as the reader who persists to Chapter 18 will believe.

EXERCISES

1. Verify directly the naturality formula for $\bar{\theta}_f$.
2. Interpret the Hopf invariant $H: \pi_{2n-1}(S^n) \to Z$ as a (non-stable) functional cohomology operation.
3. Let $K = S^n \cup_f e^{n+k}$ for some $f: S^{n+k-1} \to S^n$. Suppose there exists a secondary cohomology operation Φ defined and non-zero on the generator of $H^n(K)$. Then prove f is essential. (We say Φ detects f.)
4. Let $\bar{u} \in H^p(X)$, $\bar{v} \in H^q(X)$, $\bar{w} \in H^r(X)$ (all with integer coefficients) have the properties $\bar{u}\bar{v} = 0$ and $\bar{v}\bar{w} = 0$. Let u, v and w be cochain representatives, and let a, b be such that $u \cup v = \delta a$ and $v \cup w = \delta b$. Prove:
   i. $u \cup b + (-1)^{p+1}(a \cup w)$ is a cocycle
   ii. The cocycle of (i) projects to an element of the quotient group

   $$H^{p+q+r-1}(X)/(H^{p+q-1}(X)\bar{w} + \bar{u}H^{q+r-1}(X))$$

   which depends only on $\bar{u}$, $\bar{v}$ and $\bar{w}$

   This element is denoted $\langle \bar{u}, \bar{v}, \bar{w} \rangle$ and is called the *Massey triple product* of the three cohomology classes.

REFERENCES

Functional cohomology operations
1. F. P. Peterson and N. Stein [1].
2. N. E. Steenrod [4].

Secondary operations
1. J. F. Adams [2].
2. H. Cartan [6].

Higher-order operations
1. C. R. F. Maunder [1].

Functional higher-order operations
1. F. P. Peterson [1].

CHAPTER 17

COMPOSITIONS IN THE STABLE HOMOTOPY OF SPHERES

In this chapter we will take a closer look at the stable homotopy groups of spheres in the range computed in Chapter 12. As an example of the sort of question we ask, consider the Hopf map $\eta: S^3 \to S^2$. Under suspension, η gives rise to a map $E^n\eta: S^{n+3} \to S^{n+2}$ for each positive integer n. This element, also denoted by η, is the generator of the stable 1-stem. We have seen that the stable 2-stem is a cyclic group of order 2. Is it generated by the composition $E^{n-2}\eta \circ E^{n-1}\eta: S^{n+2} \to S^n$, denoted η^2, or is this composition null-homotopic?

We will study such compositions as η^2 as well as compositions of a somewhat different nature, the so-called secondary compositions, which we now introduce.

SECONDARY COMPOSITIONS

Suppose we are given homotopy classes $\beta \in [X, Y]$ and $\alpha \in [Y, Z]$, as shown,

$$\begin{array}{ccccc} X & \xrightarrow{\beta} & Y & \xrightarrow{\alpha} & Z \\ & & {\scriptstyle i}\downarrow & \nearrow {\scriptstyle \bar{\alpha}} & \\ & & K = Y \cup_\beta CX & & \end{array}$$

and we wish to extend α (or rather a representative of α) over $K = Y \cup_\beta CX$. It is an easy exercise to verify directly that such an extension can be found if and only if the composite $\alpha \circ \beta$ is inessential, that is, if $\alpha \circ \beta = 0 \in [X,Z]$. In fact, we have an exact sequence

$$[X,Z] \xleftarrow{\beta^*} [Y,Z] \xleftarrow{i^*} [Y \cup_\beta CX, Z] \xleftarrow{j^*} [SX,Z]$$

as in Chapter 14. This shows that α can be extended over all of $K = Y \cup_\beta CX$, that is, α is in im $(i^\#)$, if and only if α is in ker $(\beta^\#)$, that is, $\alpha \circ \beta = 0$.

If $\bar{\alpha}$ is a homotopy class such that $i^\#(\bar{\alpha}) = \alpha$, we call $\bar{\alpha}$ an *extension* of α. Of course, this is consistent with the usual sense of the word "extension."

Working in the stable range, we may assume that all the sets of homotopy classes under discussion have natural abelian group structure. Then it is clear from the above exact sequence that any two extensions (in the present sense) of α differ by the image under $j^\#$ of a map of SX into Z. We should make explicit that the cone $CX \subset K$ is the "upper" cone, i.e., is a quotient space of $X \times [0,1]$, while SX is represented by $CY \cup K$, where CY is the "lower" cone, i.e., a quotient space of $Y \times [-1,0]$, with $Y \times 0 \subset CY$ identified in the natural way with $Y \subset K$. (We have remarked in Chapter 14 that SX and $CY \cup K$ have the same homotopy type.)

We will now introduce a kind of companion or dual to the notion of extension as defined above. Let γ be a homotopy class of maps $W \to X$. Then a representative of γ induces a map $CW \to Y \cup_\beta CX = K$ (where CW is the "upper" cone) in an obvious way. We ask when a map $C\gamma: CW \to K$ admits an extension to a map

$$\tilde{\gamma}: (SW, C_-W) \to (K, Y)$$

that is, to an extension of $C\gamma$ over all of SW with the property that the extension $\tilde{\gamma}$ maps all of the "lower" cone of SW into $Y \subset K$. It is easy to see directly that this is possible if and only if the composite $\beta \circ \gamma: W \to Y$ is null-homotopic. When this occurs, we say that $\tilde{\gamma}$ is a *co-extension* of γ.

Note that if we take any co-extension $\tilde{\gamma}$ and follow it by the identification map $K \to SX$ obtained by collapsing $Y \subset K$ to a point, we obtain $S\gamma: SW \to SX$. (We could just as easily call this map $E\gamma$ instead of $S\gamma$. Both notations are in widespread use.)

The choice of co-extension is seen to be equivalent to the choice of the null-homotopy of $\beta \circ \gamma: W \to Y$. Thus in the stable range any two co-extensions of γ differ by a map of $SW \to Y$, since such a map represents in a natural way the difference between two null-homotopies of $\beta \circ \gamma$.

Now suppose we are given homotopy classes

$$W \xrightarrow{\gamma} X \xrightarrow{\beta} Y \xrightarrow{\alpha} Z$$

such that both $\beta \circ \gamma$ and $\alpha \circ \beta$ are zero. Then we can form an extension $\bar{\alpha}$ of α and a co-extension $\tilde{\gamma}$ of γ. The composite defines a map

$$SW \xrightarrow{\tilde{\gamma}} K = Y \cup_\beta CX \xrightarrow{\bar{\alpha}} Z$$

and it remains to see what the indeterminacy is; that is, the above composite varies with the choice of $\bar{\alpha}$ and $\tilde{\gamma}$. In the diagram below, the unmarked

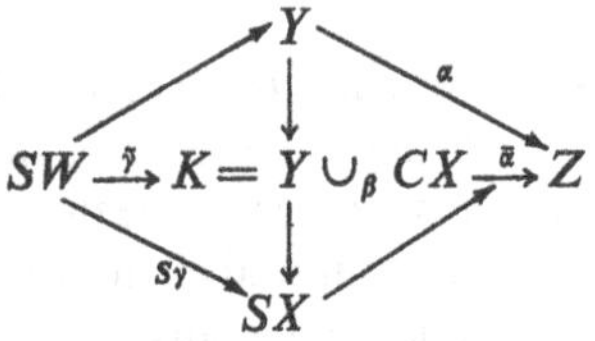

arrow $SX \to Z$ represents the indeterminacy in the choice of $\bar{\alpha}$, while the unmarked arrow $SW \to Y$ represents the indeterminacy in the choice of $\tilde{\gamma}$. Recalling the assumption of stability, the diagram makes obvious that the indeterminacy of $\bar{\alpha} \circ \tilde{\gamma}$ is $\alpha_{\#}[SW, Y] + (S\gamma)^{\#}[SX,Z]$. Thus the above composition $\bar{\alpha} \circ \tilde{\gamma}$: $SW \to Z$ defines the *secondary composition* or *Toda bracket* $\langle\alpha,\beta,\gamma\rangle$ as a uniquely determined element of the quotient group $[SW,Z]/Q$ where $Q = \alpha_{\#}[SW, Y] + (S\gamma)^{\#}[SX,Z]$.

For heuristic purposes we can appeal to Figure 1. The co-extension $\tilde{\gamma}$ maps the upper cone of SW by $S\gamma$ into $CX \subset K$, maps the "equatorial" W by $\beta \circ \gamma$, and maps the lower cone of SW by the null-homotopy of $\beta \circ \gamma$. The extension $\bar{\alpha}$ maps Y into Z by α and maps $CX \subset K$ into Z by the null-homotopy of $\alpha \circ \beta$.

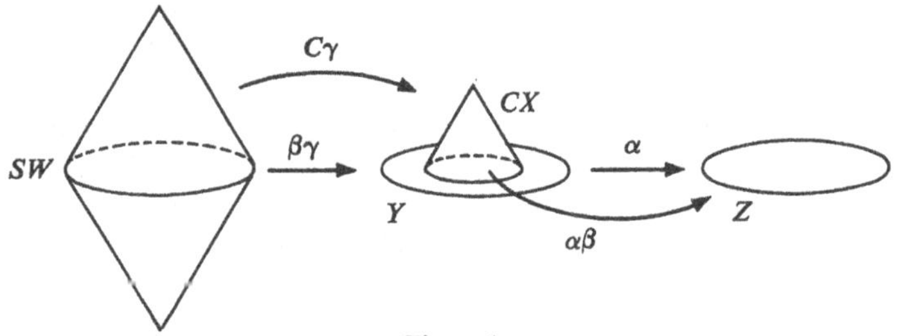

Figure 1

Now suppose W, X, Y, and Z are spheres of dimensions sufficiently large so that α, β, and γ are in the stable range. We would then expect the stable class of $\langle\alpha,\beta,\gamma\rangle$ to depend only on α, β, and γ viewed as elements of G and not on the dimensions of the chosen representatives.

While this is not quite true, it is true up to sign, and we may define a stable secondary composition as follows. Let $\alpha \in G_h$, $\beta \in G_k$, and $\gamma \in G_j$. Suppose $\alpha \circ \beta$ and $\beta \circ \gamma$ are zero. Let a, b, and c represent α, β, and γ as follows:

$$S^{n+h+k+j} \xrightarrow{c} S^{n+h+k} \xrightarrow{b} S^{n+h} \xrightarrow{a} S^n$$

Then the class of $(-1)^{n-1}\langle a,b,c\rangle$ in $G_{h+k+j+1}/(\alpha \circ G_{k+j+1} + G_{h+k+1} \circ \beta)$ is independent of n and defines the stable secondary composition $\langle\alpha,\beta,\gamma\rangle$.

THE 0-, 1-, AND 2-STEMS

We now begin our detailed analysis of the stable k-stem, G_k, that is, the group $\pi_{n+k}(S^n)$ for k small and n large (and of course working in the 2-component only).

We have in the 0-stem an infinite cyclic group generated by the identity map $\iota\colon S^n \to S^n$. We may say ι is detected by its induced cohomology homomorphism ι^*, since $\iota^*\colon H^n(S^n) \to H^n(S^n)$ is non-zero. We also observe that 2ι is detected by Sq^1; that is, $Sq^1(\sigma^n) = e^{n+1}$ in the complex $S^n \cup_{2\iota} e^{n+1}$ and the functional operation $Sq^1_{2\iota}$ is non-trivial. This result may be generalized as follows.

Proposition 1

$2^r\iota$ is detected by the Bockstein operator d_r, which is non-zero in $S^n \cup_{2^r\iota} e^{n+1}$.

The proof is left as an exercise.

The stable 1-stem G_1 is the cyclic group Z_2 generated by the Hopf map η, detected by Sq^2.

We recall that the stable 2-stem G_2 is also cyclic of order 2; the next result identifies a generator and answers the question posed at the beginning of this chapter.

Proposition 2

The composition η^2 is non-zero and hence generates G_2.

We shall give two proofs of this.

FIRST PROOF: We will derive a contradiction from the assumption that η^2 is null-homotopic.

Suppose then that $\eta^2 = 0$. For

$$W \xrightarrow{\gamma} X \xrightarrow{\beta} Y$$

we take

$$S^{n+2} \xrightarrow{\eta} S^{n+1} \xrightarrow{\eta} S^n$$

so that $K = Y \cup_\beta CX$ becomes $K = S^n \cup_\eta e^{n+2}$. We form a co-extension

$$\tilde{\eta}: S^{n+3} \to K$$

and consider the complex $L = K \cup_{\tilde{\eta}} e^{n+4}$.

For simplicity we use the symbol e^k to denote not only the cell in question, as a cell of K or L, but also to denote the corresponding generator of $H^k(K)$ or $H^k(L)$. In general cohomology will be understood to have coefficients in Z_2.

Now the Hopf map η has the property that $Sq^2(\sigma^n) = e^{n+2}$ in the complex K. The same relation must hold in L, by naturality of Sq^2 with respect to the inclusion i: $K \subset L$, since i^* is isomorphic in dimensions n and $n+2$.

We have observed that $p\tilde{\eta} = S\eta$ where p: $K \to S^{n+2}$ is the map which pinches S^n to a point. Thus the diagram below is commutative, where we

$$\begin{array}{ccc} S^{n+3} & \xrightarrow{\tilde{\eta}} & K \\ \Big\downarrow = & & \Big\downarrow p \\ S^{n+3} & \xrightarrow[\eta]{} & S^{n+2} \end{array}$$

have written η for $S\eta$ as usual. Now we verified in the last chapter that Sq^2_η: $H^{n+2}(S^{n+2}) \to H^{n+3}(S^{n+3})$ is non-zero, with zero indeterminacy. Applying the naturality formula for functional operations to the above commutative diagram, we deduce that $Sq^2_{\tilde{\eta}}(e^{n+2}) = e^{n+3}$, again with zero indeterminacy. (Here we might be precise and identify e^{n+2} and e^{n+3} as the generators of $H^{n+2}(K)$ and $H^{n+3}(S^{n+3})$, respectively; but this is clear from the context, and we will generally leave these matters implicit.)

Now if $Sq^2_{\tilde{\eta}}(e^{n+2}) = e^{n+3}$, then $Sq^2(e^{n+2})$ must be e^{n+4} in the complex L. To see this, recall that $Sq^2_{\tilde{\eta}}$ is defined by the following diagram, since L is

$$\begin{array}{ccc} & & H^{n+2}(L) \to H^{n+2}(K) \\ & & \Big\downarrow Sq^2 \\ H^{n+3}(S^{n+3}) & \to & H^{n+4}(L) \end{array}$$

precisely the space obtained by converting $\tilde{\eta}$ to an inclusion $\tilde{\eta}$: $S^{n+3} \to K \cup_{\tilde{\eta}} (S^{n+3} \times I)$ and then pinching the subspace identified with S^{n+3} to a point, and thus L plays the role of K/S^{n+3} (or of the pair (K, S^{n+3})).

We have thus shown that, in the complex L,

$$Sq^2 Sq^2(\sigma^n) = Sq^2(e^{n+2}) = e^{n+4}$$

so that the composition $Sq^2 Sq^2$ is non-zero. But this is impossible, in the light of the Adem relation $Sq^2 Sq^2 = Sq^3 Sq^1$. Obviously $Sq^1(\sigma^n) = 0$; in the first place, σ^n is the reduction (mod 2) of a class with integer coefficients and

Sq^1 is the Bockstein; in the second place, there is nothing that $Sq^1(\sigma^n)$ could possibly be, since $H^{n+1}(L)=0$.

Thus the assumption that $\eta^2=0$ leads to a contradiction based on the hypothetical co-extension $\tilde{\eta}$. This proves the proposition.

It is fruitful to think of the contradiction as arising from the relation $Sq^2Sq^2=0$, which holds on any class which is in the image of the reduction map from integral coefficients to Z_2 coefficients.

The above proof could be summarized as follows: Suppose $\eta^2=0$; then we can form a co-extension $\tilde{\eta}: S^{n+3}\to K=S^n\cup_\eta e^{n+2}$ and form the complex $L=K\cup_{\tilde{\eta}} e^{n+4}$. By the naturality formula for functional operations, $Sq^2_{\tilde{\eta}}(e^{n+2})=e^{n+3}$ (with zero indeterminacy,) and therefore $Sq^2(e^{n+2})=e^{n+4}$ in L. Thus Sq^2Sq^2 is non-zero in L, which is impossible. Thus η^2 is non-zero.

We have given the above proof in complete detail; future proofs will look more like the above summary than like the full-length proof which preceded it.

SECOND PROOF: To illustrate another technique, we give a second proof, using the first Peterson-Stein formula. We will use the complexes $K=S^{n+1}\cup_\eta e^{n+3}$ and $L=S^n\cup_{\eta^2} e^{n+3}$ and the map $f: K\to L$ which is given by $\eta: S^{n+1}\to S^n$ and by the identity on the interior of e^{n+3}.

Since such maps are used frequently, it might be well to be precise on this occasion and explain exactly how f is defined; then, in the sequel, the brief description given above will be considered sufficient. Recall that K is obtained from the disjoint union $K_0=S^{n+1}\cup e^{n+3}$ by identifying certain points; under the identification map $p: K_0\to K$, $p(x)=p(y)$ if $\eta(y)=x$, where $y\in S^{n+2}=bdy(e^{n+3})$. Similarly L is the identification space obtained from the disjoint union $L_0=S^n\cup e^{n+3}$ and the map $q: L_0\to L$ such that $q(x)=q(y)$ if $\eta^2(y)=x$, where $y\in S^{n+2}=bdy(e^{n+3})$. We have an obvious map $f_0: K_0\to L_0$ given by η and the identity map of e^{n+3}. We consider the composite $qf_0: K_0\to L$. This map is constant on each p-equivalence class of points of K_0, i.e., the diagram below, considered as a diagram of

$$\begin{array}{ccc} K_0 & \xrightarrow{f_0} & L_0 \\ {\scriptstyle p}\downarrow & & \downarrow{\scriptstyle q} \\ K & \xrightarrow{f} & L \end{array}$$

sets and transformations of sets, may be completed (in a unique way) by a transformation f, to yield a commutative diagram. It follows that f is a continuous map (this is the fundamental property of the identification topology of K), and this is the map f which we will use.

We now give a second proof that η^2 is essential.

Let K, L, and $f: K \to L$ be as just above. In the commutative diagram below, where the vertical maps are the inclusions, we have $Sq^2_\eta(\sigma^n) = \sigma^{n+1}$

$$\begin{array}{ccc} K & \xrightarrow{f} & L \\ \downarrow & & \downarrow \\ S^{n+1} & \xrightarrow{\eta} & S^n \end{array}$$

(with zero indeterminacy) and therefore, by naturality, $Sq^2_f(\sigma^n) = \sigma^{n+1}$ (with zero indeterminacy), where in this last formula σ^n and σ^{n+1} denote cohomology classes of L and K, respectively. Now we can define a secondary operation Φ based on the following diagram, which is part of the

$$\begin{array}{ccccc} F = K(Z_2, n+1) \xrightarrow{i} & X_{n+1} & \xrightarrow{\varphi} K(Z_2, n+3) & & \\ & \downarrow p & & & \\ B = & K(Z,n) & \xrightarrow{\theta} K(Z_2, n+2) & \xrightarrow{\psi} & K(Z_2, n+4) \end{array}$$

Postnikov system for S^n, with $\theta = Sq^2$, $\psi = Sq^2$, and φ equal to the class α of Chapter 12. Then, by the first Peterson-Stein formula (Theorem 1 of Chapter 16),

$$f^*(\Phi(\sigma^n)) = Sq^2(Sq^2_f(\sigma^n)) = Sq^2(\sigma^{n+1}) = e^{n+3}$$

with indeterminacy $Sq^2(f^*(H^{n+1}(L))) = 0$. But if $f^*(\Phi(\sigma^n)) = e^{n+3}$, then $\Phi(\sigma^n) = e^{n+3} \in H^{n+3}(L)$, also with zero indeterminacy. However, if $L = S^n \cup_{\eta^2} e^{n+3}$ has this non-zero secondary operation $\Phi(\sigma^n) = e^{n+3}$, then the attaching map η^2 cannot be null-homotopic (see Exercise 3 of Chapter 16).

Notice again that it is the relation $Sq^2Sq^2 = 0$ (on classes with integer coefficients) that provides the key step, and in fact we have proved the following corollary.

Corollary 1

Let Φ be the secondary operation associated with the relation $Sq^2Sq^2 = 0$ (valid on cohomology with integer coefficients). Then Φ detects η^2.

There is another question which we may ask about the 2-stem. Since the 1-stem is of order 2, the composition of η and 2ι is null-homotopic whichever way we form it; thus the diagram

$$S^{n+1} \xrightarrow{2\iota} S^{n+1} \xrightarrow{\eta} S^n \xrightarrow{2\iota} S^n$$

gives rise to a secondary composition $\langle 2\iota, \eta, 2\iota \rangle$. This bracket takes its value in a quotient group of $[S(S^{n+1}), S^n] = \pi_{n+2}(S^n)$, and in fact the indeterminacy is readily seen to be

$$(2\iota)_\#(\pi_{n+2}(S^n)) + (2\iota)^\#(\pi_{n+2}(S^n))$$

which is zero, since the 2-stem is a group of order 2. Thus the bracket $\langle 2\iota,\eta,2\iota\rangle$ must be either η^2 or zero. We can settle this question by a direct argument.

Proposition 3

$$\langle 2\iota,\eta,2\iota\rangle = \eta^2 \in \pi_{n+2}(S^n).$$

PROOF: Let f represent the bracket, and let $K = S^n \cup_f e^{n+3}$. Let Ψ denote the secondary operation associated with the relation $Sq^2Sq^2 + Sq^3Sq^1 = 0$; this operation was described in some detail in the last chapter. We will show that Ψ is non-zero in K, which will imply that f is essential, from which the result follows.

Let $L = S^n \cup_\eta e^{n+2}$, and let $\widetilde{2\iota}$ be a co-extension of the map 2ι: $S^{n+1} \to S^{n+1}$. Then $\widetilde{2\iota}$: $S^{n+2} \to L$ and we can form the complex $M = L \cup_{\widetilde{2\iota}} e^{n+3}$. Now $Sq^3 = Sq^1Sq^2$ is non-zero in M; in fact this is the essential property of M for our purposes. It is clear that $Sq^2(\sigma^n_M) = e^{n+2}$, and the fact that $Sq^1(e^{n+2}) = e^{n+3}$ follows by naturality from the obvious fact that Sq^1 is non-zero in $M/S^n = S^{n+2} \cup_{2\iota} e^{n+3}$.

We now map M into K by a map g which takes $L \subset M$ into S^n by the extension $\overline{2\iota}$, maps the boundary of e^{n+3} by f, and maps the rest of $e^{n+3} = CS^{n+2}$ by Cf.

Since g maps $S^n \subset M$ to $S^n \subset K$ by 2ι, we have $Sq^1_g(\sigma^n_K) = \sigma^n_M$. On the other hand, Sq^2_g is clearly zero.

By the first Peterson-Stein formula,

$$\begin{aligned} g^*(\Psi(\sigma^n_K)) &= Sq^2(Sq^2_g(\sigma^n_K)) + Sq^3(Sq^1_g(\sigma^n_K)) \\ &= 0 + Sq^3(\sigma^n_M) \\ &= e^{n+3}_M \end{aligned}$$

and the indeterminacy is zero; thus $\Psi(\sigma^n_K) = e^{n+3}_K$ with zero indeterminacy, and this completes the proof.

Corollary 2

Let Ψ be the secondary operation associated with the relation $Sq^2Sq^2 + Sq^3Sq^1 = 0$. Then Ψ detects η^2.

It is instructive to compare the above operation Ψ with the operation Φ used previously. We may view Φ as defined on classes in $H^n(\;;Z_2)$ which are in the image of reduction mod 2 from integral cohomology and which are in the kernel of Sq^2. The operation Ψ is defined on a larger subset but with larger indeterminacy. Whenever both operations are defined, they coincide, modulo the larger indeterminacy. In the particular case of

$K = S^n \cup_{\eta^2} e^{n+3}$, we see that both Φ and Ψ are defined and non-zero modulo zero.

We sketch an alternate proof that $\langle 2\iota, \eta, 2\iota \rangle = \eta^2$. Suppose that $\overline{2\iota} \circ \widetilde{2\iota}$ is null-homotopic; then we can form a co-extension $\widetilde{\widetilde{2\iota}}$ of $\overline{2\iota}$. This will be a map of S^{n+3} into $S^n \cup_{\overline{2\iota}} C(S^n \cup_\eta e^{n+2})$. In this latter complex, $Sq^2 Sq^1$ is non-zero. If we now attach e^{n+4} by $\widetilde{\widetilde{2\iota}}$, $Sq^1(Sq^2 Sq^1)$ is non-zero, but $Sq^2 Sq^2$ is necessarily zero, which contradicts the Adem relations.

THE 3-STEM

We now turn our attention to G_3, the stable 3-stem. We have seen that the 2-primary part of this group is Z_8, and it is natural to conjecture that the Hopf map $\nu: S^{n+3} \to S^n$ is a generator.

Proposition 4

The Hopf map ν generates the 2-primary component of G_3.

PROOF: By the suspension theorem (Theorem 1 of Chapter 12), the group $\pi_8(S^5)$ is the stable group Z_8 and the suspension homomorphism $E: \pi_7(S^4) \to \pi_8(S^5)$ is an epimorphism. It thus suffices to prove that $E\nu$ is not divisible by 2 in Z_8.

Consider the Hopf invariant homomorphism $H: \pi_7(S^4) \to Z$. If $f: S^7 \to S^4$ then $H(f)$, reduced mod 2, is nothing other than $Sq_f^4(1)$, but Sq_f^4 commutes with suspension. Thus every element in the kernel of the suspension E has even Hopf invariant.

Suppose $E\nu$ is divisible by 2; say, $E\nu = 2\alpha$. Since E is an epimorphism, we may write $\alpha = E\beta$. Thus $E(\nu - 2\beta) = 0$. Thus $\nu - 2\beta$ has an even Hopf invariant, which is impossible, since H is a homomorphism and ν has an odd Hopf invariant.

We next consider the composition η^3 in G_3. To study η^3, we will use the following relation between secondary operations. Let K be a complex and let u be an integral class, $u \in H^n(K;Z)$, such that $Sq^2 u = 0$. Let Φ be the secondary operation based on the relation $Sq^2 Sq^2 = 0$ for integral classes (Φ detects η^2).

Lemma 1

$Sq^2(\Phi(u)) = d_2(Sq^4(u)) \in H^{n+5}(K;Z_2)/Sq^1 H^{n+4}(K;Z_2)$.

PROOF: It is quite natural to look for this relation, since, in the calculations (Chapter 12) of the Postnikov system of S^n, we found that $d_2 Sq^4(\iota_n)$ vanished in $H^*(X_2)$ because it was equal to $\tau(Sq^2(\iota_{n+2})) = Sq^2(\alpha)$ where α is the class which gives rise to Φ.

This will serve as the universal example for the proof; taking $K = X_1$ and $u = p = \iota_n \in H^n(X_1;Z)$, we have $Sq^2u = 0$, and by our calculations (Chapter 12) we have

$$Sq^2(\Phi(\iota_n)) = d_2Sq^4(\iota_n)$$

with zero indeterminacy.

Now in the general case, $Sq^2u = 0$; so we can find a lifting $v: K \to X_1$ such that $pv = u$, which is to say that $u = v^*(p)$. The lemma follows by naturality. The total indeterminacy would be

$$Sq^2(Sq^2H^{n+1}(K;Z_2)) + Sq^1(H^{n+4}(K;Z_2))$$

but the first term is contained in the second, since $Sq^2 = Sq^2\,Sq^1(Sq^2Sq^1)$. (When $K = X_1$, the fact that $Sq^1H^{n+4} = 0$ follows from the calculations of Chapter 12.)

We can now evaluate the composition η^3.

Proposition 5

In the stable 3-stem G_3, $\eta^3 = 4\nu$.

PROOF: If $\eta^2 \circ \eta$ were null-homotopic, then we could form a co-extension $\tilde{\eta}$: $S^{n+4} \to K = S^n \cup_{\eta^2} e^{n+3}$ and form the complex $L = K \cup_{\tilde{\eta}} e^{n+5}$. We have been before that the operation Φ of the lemma detects η^2 and $\Phi(\sigma_L^n) = e_L^{n+3}$ with zero indeterminacy. On the other hand, $Sq^2(e_L^{n+3}) = e^{n+5}$, since L/S^n is just $S^{n+3} \cup_\eta e^{n+5}$ and here Sq^2 is certainly non-zero. Thus we have shown that in L, $Sq^2(\Phi(\sigma^n)) = e^{n+5}$. But $Sq^4(\sigma^n)$ is obviously zero, and $d_2Sq^4(\sigma^n) = 0$ with zero indeterminacy. This contradicts Lemma 1. Therefore η^3 is essential.

But η is of order 2, since the 1-stem is Z_2, and therefore η^3 is also of order 2. The 2-component of the 3-stem is Z_8, and the only non-zero element of order 2 in this group is 4ν. Therefore η^3 must be 4ν. This completes the proof.

The reader may well suspect that the relation $Sq^2\Phi = d_2Sq^4$ between secondary operations gives rise to a tertiary operation and that this tertiary operation detects 4ν. This suspicion is indeed correct, but we do not pursue it.

We can form the bracket $\langle \eta, 2\iota, \eta \rangle$, since $2\eta = 0$. We conclude our discussion of the 3-stem by evaluating this bracket.

Proposition 6

$\langle \eta, 2\iota, \eta \rangle = 2\nu$ modulo 4ν.

The indeterminacy of this bracket is the subgroup of Z_8 generated by $4\nu = \eta^3$. Thus the proposition asserts that the bracket contains precisely the elements 2ν and 6ν.

To prove the proposition, suppose for some value of r, $0 \leq r < 3$, we have $2^r\nu \in \langle \eta, 2\iota, \eta \rangle$. Recall that the bracket is the composition

$$S^{n+3} \xrightarrow{\tilde{\eta}} K = S^{n+1} \cup_{2\iota} e^{n+2} \xrightarrow{\bar{\eta}} S^n$$

and consider the composition

$$S^{n+3} \xrightarrow{2^r\iota \vee \tilde{\eta}} S^{n+3} \vee K \xrightarrow{\nu \vee \bar{\eta}} S^n$$

We are assuming this composition null-homotopic. Using an extension of $\nu \vee \bar{\eta}$ as attaching map, we may construct a complex

$$L = S^n \cup e^{n+2} \cup e^{n+3} \cup e^{n+4} \cup e^{n+5}$$

The following relations hold in the cohomology of L: $Sq^2\sigma^n = e^{n+2}$; $Sq^1 e^{n+2} = e^{n+3}$; $Sq^2 e^{n+3} = e^{n+5}$; $Sq^4\sigma^n = e^{n+4}$; and $d_r e^{n+4} = e^{n+5}$. From the Adem relation $Sq^2Sq^1Sq^2 = Sq^1Sq^4 + Sq^4Sq^1 = 0$, it follows that $r = 1$. This completes the proof.

We remark that we may use the above relation, written for example as $Sq^1Sq^4 + Sq^4Sq^1 + (Sq^2Sq^1)Sq^2 = 0$, to define a secondary operation Θ. The reader may easily verify that Θ detects 2ν.

Note also the hierarchy which has appeared in the 3-stem: ν is detected by the primary operation Sq^4; 2ν, by the secondary operation Θ; and 4ν, by a tertiary operation. A similar phenomenon appears in the 0-stem—we may view d_r as an r-ary operation and even view ι^* as a 0-ary operation (the induced homomorphism).

THE 6- AND 7-STEMS

We next consider G_6, where the 2-component is Z_2. It is natural to ask whether the generator is ν^2.

Proposition 7

ν^2 is essential and hence generates the 2-component of G_6.

PROOF: In $H^*(K(Z,n))$, $Sq^4Sq^4\iota_n = Sq^6Sq^2\iota_n$. Let Ξ denote the associated secondary operation. We define a map f from $S^{n+3} \cup_\nu e^{n+7}$ to $S^n \cup_{\nu^2} e^{n+7}$ using ν: $S^{n+3} \to S^n$. Then, by the first Peterson-Stein formula,

$$\begin{aligned} f^*(\Xi(\sigma^n)) &= Sq^4(Sq_f^4(\sigma^n)) + Sq^6(Sq_f^2(\sigma^n)) \\ &= e^{n+7} + 0 \end{aligned}$$

(with zero indeterminacy), and thus Ξ is non-zero in $S^n \cup_{\nu^2} e^{n+7}$, which proves that ν^2 is essential.

For another proof, see Exercise 2.

We now state the value of a secondary composition in the 6-stem.

Proposition 8

$\langle\eta,\nu,\eta\rangle = \nu^2$ mod zero.

The proof is left to the reader as Exercise 3; it is of course sufficient to show that the bracket is nonzero.

We turn next to the 7-stem. According to the results of Chapter 12, G_7 is cyclic of order 16 (in the 2-component). In a manner analogous to Proposition 4 and by a similar proof which we omit, we have the following.

Proposition 9

The Hopf map σ generates the 2-component of G_7.

Thus the generator of the 7-stem is detected by the primary operation Sq^8. Further, a hierarchy similar to that in the 3-stem occurs; namely, 2σ can be detected by a secondary operation; 4σ, by a tertiary; and 8σ, by a quaternary operation. The only secondary composition we will consider in the 7-stem is the following.

Proposition 10

$\langle\nu,8\iota,\nu\rangle = 8\sigma$ modulo zero.

That the indeterminacy of the bracket vanishes is obvious for dimensional reasons. The evaluation of the bracket is similar to the proof of Proposition 6; however, we first need a lemma, valid for n sufficiently large.

Lemma 2

Let u be an n-dimensional cohomology class—that is, let $u \in H^n(X;Z)$ for some space X. Suppose Sq^2u and $\Phi(u)$ are zero. Then $Sq^4d_3Sq^4u = d_3Sq^8u$ modulo the total indeterminacy.

PROOF: In the calculations of Chapter 12, we have proved this lemma in the special case $u = \iota_n$ in $H^n(X_2;Z)$ (see the table on page 117 and Lemma 2 of Chapter 12). But this is the universal example; the general case follows by naturality.

PROOF OF PROPOSITION 10: Proceeding as in the proof of Proposition 6, suppose we have $2^r\sigma$ in $\langle\nu,8\iota,\nu\rangle$ for some r, $0 \leq r < 4$. We then may construct a complex

$$L = S^n \cup e^{n+4} \cup e^{n+5} \cup e^{n+8} \cup e^{n+9}$$

with the following properties: $Sq^4\sigma^n = e^{n+4}$; $d_3e^{n+4} = e^{n+5}$; $Sq^4e^{n+5} = e^{n+9}$; $Sq^8\sigma^n = e^{n+8}$; and $d_re^{n+8} = e^{n+9}$. By Lemma 2 it follows that $r=3$. This completes the proof.

Of course, a quaternary operation which detects 8σ may be based on the relation of Lemma 2.

We summarize the results of our computations in an omnibus theorem.

Theorem 1

The 2-components of the groups G_k, $k \leq 7$, are given by the table below.

STEM	GROUP	GENERATOR AND RELATIONS
0	Z	ι
1	Z_2	η
2	Z_2	$\eta^2 = \langle 2\iota,\eta,2\iota\rangle$
3	Z_8	ν; $2\nu = \langle \eta,2\iota,\eta\rangle \bmod \eta^3 = 4\nu$
4	0	
5	0	
6	Z_2	$\nu^2 = \langle \eta,\nu,\eta\rangle$
7	Z_{16}	σ; $\langle \nu,8\iota,\nu\rangle = 8\sigma$

DISCUSSION

We have considered only the simplest sort of Toda bracket. It is not possible to express 2σ as a bracket without considerably complicating the allowable construction.

However, we have evaluated many important compositions and secondary compositions; in particular, we have obtained all the significant data through the stable 6-stem and much in the 7-stem. But of course we have only scratched the surface. Toda especially has taken the point of view of calculation by means of composition. In his book will be found many additional results obtained by composition methods, as well as a detailed exposition of the properties of the secondary composition.

Nor does the theory stop with *secondary* composition. Oguchi has thoroughly analyzed tertiary composition; Spanier has indicated the general framework of higher compositions.

Spanier has also shown that both the functional and secondary cohomology operations of Chapter 16 fit into the framework of secondary composition. The Peterson-Stein formulas become special cases of some general properties of the secondary composition.

Some care need be exercised with the indeterminacy, though, since the usual definition of a secondary operation has only "half" its natural indeterminacy as a secondary composition. Thus, if we have the relations $\psi\theta = 0$ and $\theta u = 0$, we might define $\Phi(u) = \langle u,\theta,\psi\rangle$, but this secondary composition has larger indeterminacy than does our definition of $\Phi(u)$. The reason for this is that our definition includes the choice of the specific cohomology class φ.

EXERCISES

1. Prove Proposition 1.
2. Give an alternative proof of Proposition 7 by imitating the first proof of Proposition 2, that is, assume ν^2 null-homotopic and form a coextension $\nu^2: S^{n+7} \to S^n \cup_\nu e^{n+4}$. Then obtain a contradiction to the Adem relation $Sq^4Sq^4 = Sq^6Sq^2 + Sq^7Sq^1$.
3. Prove Proposition 8. HINT: Imitate the alternative proof of Proposition 3.
4. Use Adem relations in similar fashion to find some non-zero stable elements in stems higher than 7.

REFERENCES

Compositions

1. F. P. Peterson [1].
2. H. Toda [1].

Secondary compositions

1. M. G. Barratt [2,3].
2. E. H. Spanier [2].
3. H. Toda [1,2].

Higher compositions

1. K. Oguchi [1].
2. E. H. Spanier [3].

CHAPTER 18

THE ADAMS SPECTRAL SEQUENCE

In Chapter 14 we constructed a spectral sequence for $[Y,X]$ based on a Postnikov decomposition of X (or a cell decomposition of Y). The spectral sequence starts at $H^*(Y;\pi_*X)$ and thus requires knowledge of the homotopy groups of X; it is not a useful tool with which to compute those homotopy groups.

In this chapter another spectral sequence, constructed by Adams, is described. This spectral sequence differs from that of Chapter 14 in the following important respects. First of all, it yields information only about the 2-component—i.e., it yields a composition series for the 2-component of $[Y,X]$. Secondly, this spectral sequence is valid only in the stable range—for example, if X is $(n-1)$-connected and dim $Y \leq 2n-2$. Thirdly, the starting point, the E_2-term, depends only on the cohomology $H^*(X;Z_2)$ and $H^*(Y;Z_2)$, viewed as graded modules over the Steenrod algebra $\mathcal{A}$.

Taking X and Y to be spheres of suitable dimensions, we would thus expect the Adams spectral sequence to give information about the 2-component of stable homotopy groups of spheres. This is in fact the case, and in recent years the Adams spectral sequence has been the most important tool for computing these groups.

RESOLUTIONS

In Chapter 12, we calculated some homotopy groups of spheres by constructing a Postnikov system. In this method of computation, the cohomology of $K(Z,n)$ is killed, one step at a time, by passing to successive fibre spaces induced by maps into $K(\pi,n)$ spaces. Among the cohomology classes of $K(Z,n)$ which had to be killed by separate maps were the classes given by Sq^2, Sq^4, and Sq^8 on the fundamental class. Since Sq^i is indecomposable whenever i is a power of 2 (Theorem 1 of Chapter 4), it is not surprising that these classes must be killed and it is natural to wonder whether it would not make sense to kill all three of these classes at the first stage of the construction. We can achieve this by a map of $K(Z,n)$ into a product space $K_2 \times K_4 \times K_8$ where $K_i = K(\pi_i, n+i)$, with an appropriate choice of the groups π_i.

As usual we will concern ourselves only with 2-primary cohomology and homotopy, and by $H^*(X)$ we mean the reduced cohomology of X with Z_2 coefficients, etc. For technical reasons, it turns out to be convenient to take all the groups π_i equal to Z_2 and moreover to start with $K(Z_2,n)$ in place of $K(Z,n)$ at the base of the tower. We therefore formulate our new kind of tower, called a complex, in the following way.

Definition

A *complex* $\mathfrak{X}$ over X, where X is a (nice) simply connected space having finitely generated homotopy groups, is a diagram (finite or infinite) of the form

$$
\begin{array}{ccccc}
 & & \vdots & & \\
 & & \big\downarrow{\scriptstyle p_{s+1}} & & \\
\Omega K_{s-1} & \xrightarrow{i_{s-1}} & X_s & \xrightarrow{g_s} & K_s \\
 & & \big\downarrow{\scriptstyle p_s} & & \\
 & & \vdots & & \\
\Omega K_0 & \xrightarrow{i_0} & X_1 & \xrightarrow{g_1} & K_1 \\
 & & \big\downarrow{\scriptstyle p_1} & & \\
 & & X_0 = X & \xrightarrow{g_0} & K_0
\end{array}
$$

where each K_s is a generalized Eilenberg-MacLane space

$$K_s = \prod_{j=1}^{r_s} K_{s,j} \qquad K_{s,j} = K(Z_2, n_{s,j})$$

and each X_{s+1} is the (nice) space induced over X_s by the map g_s from the standard contractible fibring over K_s.

A complex $\mathfrak{X}$ will be called *$(n-1)$-connected* if every X_s is $(n-1)$-connected $(s \geq 0)$ and $n_{s,j} \geq n$ for all (s,j).

The case $X = S^n$, in which one might expect to collect into the map g_0 all the indecomposable squares, suggests that the rank r_s should be allowed to be infinite. On the other hand, we will be studying $[Y,X]$ for dim Y finite and bounded by a bound depending on the connectivity of X. Thus we can assume that the rank r_s is finite for every s, without much loss of generality, and at considerable gain in technical simplicity.

Recall that in the Postnikov system for the sphere we endeavored to kill *all* the cohomology of X by the successive choices of the k-invariants. This idea underlies the following notion.

Definition

A complex $\mathfrak{X}$ over X is *acyclic* if the induced map g_s^* maps $H^*(K_s)$ *onto* $H^*(X_s)$ for every $s \geq 0$.

An acyclic complex over X is called a *resolution of* X.

However, it will be sufficient to study complexes which are acyclic "up to a certain dimension." It is therefore convenient to have the notion of an *N-acyclic complex*, by which we mean a complex for which g_s^k is onto $H^k(X_s)$ for every s and for all k in the range $0 \leq k \leq N$. Such a complex may be called an *N-resolution.*

Before giving a characterization of acyclic complexes in terms of the induced maps p_s^*, we make some observations on the cohomology of K_s. We know by Theorem 2 of Chapter 9 that $H^*(K_{s,j})$ looks exactly like the Steenrod algebra $\mathcal{A}$ (applied to the fundamental class) in the range from dimension $n_{s,j}$ to dimension $2(n_{s,j})$. We may therefore assert that, in an $(n-1)$-connected complex, through dimensions $\leq 2n-1$ $H^*(K_s)$ is isomorphic to the free $\mathcal{A}$-module with the fundamental classes $\{\iota_{s,j}\}$ as basis.

Proposition 1

Let $N \leq 2n-1$. Then an $(n-1)$-connected complex is N-acyclic if and only if p_s^k is the zero homomorphism for every $s \geq 1$ and $k \leq N$.

We will usually formulate such propositions as the above without giving details about the range of dimensions for which the assertions are valid; instead, we will say "in the stable range" and leave the precise interpretation to the reader. For example, Proposition 1 in this form reads: In the stable range a complex is acyclic if and only if p_s^k is the zero homomorphism for every $s \geq 1$.

The proof of Proposition 1 is easy. If the complex is acyclic and we have a class $u \in H^k(X_s)$, then $u = g_s^k(v)$ for some $v \in H^k(K_s)$. But in the stable range, $H^*(K_s)$ is a free $\mathcal{A}$-module, and since the squares commute with the transgression, $u = \tau({}^1v)$, ${}^1v \in H^{k-1}(\Omega K_s)$. It follows that $p_{s+1}^k(u) = 0$, by Serre's exact sequence. Thus p^* is zero. The argument can be reversed to obtain the converse.

We will often apply the functor $[Y,\ ]$ (where Y is a nice space) to the objects and maps of a complex. Recall that $[Y,K(Z_2,n)] = H^n(Y)$, and observe that we obtain a similar classification theorem for generalized Eilenberg-MacLane spaces: $[Y,K_s] = \sum_j H^{n_{s,j}}(Y)$. In particular, if a map $Y \to K_s$ pulls all the fundamental classes back to the zero of $H^*(Y)$, then it is nullhomotopic. Note also that $2[Y,K_s] = 0$ for any Y.

By a "map of complexes" we mean the obvious thing: a family of maps $f_s\colon X_s \to Y_s$ commuting with the projections.

Proposition 2

Let $\mathcal{Y}$ be a resolution of Y and let $\mathcal{X}$ be a complex over X. Then any map $Y \to X$ can be covered by a map of complexes.

PROOF: In the diagram below we have $q^* = 0$, since Y is a resolution.

$$\begin{array}{ccccc} Y_{s+1} & \xrightarrow{f_{s+1}} & X_{s+1} & & \\ \downarrow{\scriptstyle q_{s+1}} & & \downarrow{\scriptstyle p_{s+1}} & & \\ Y_s & \xrightarrow{f_s} & X_s & \xrightarrow{g_s} & K_s \end{array}$$

Thus $q^*f_s^*g_s = 0$, which implies that the map $g_sf_sq_{s+1}\colon Y_{s+1} \to K_s$ is nullhomotopic, and this gives us the existence of a map f_{s+1} completing the diagram. Starting with $f_0 = f$, we construct the desired map of complexes by induction.

Suppose that $\mathcal{Y}$ is only an N-resolution; then if some $n_{s,j}$ is greater than N, the above argument does not go through. However, in such a context, we are only interested in the behavior of the various spaces up to dimension N, and so we lose nothing by replacing $\mathcal{X}$ by a complex in which every $n_{s,j}$ is less than or equal to N (we simply omit $K_{s,j}$ if $n_{s,j}$ is too large).

THE ADAMS FILTRATION

We now would like to use a resolution of X to study $[Y,X]$. To do so, we will need the mapping sets $[Y,X_s]$ to have natural abelian group structures. Thus to study $[Y,X]$, where X is $(n-1)$-connected, we assume $\dim Y \leq 2n-2$ and take an $(n-1)$-connected N-resolution of X, with N large.

These hypotheses of stability will not be stated at every juncture. They are there, nonetheless.

Any complex over X defines a filtration of $[Y,X]$ in a natural way. We put $F^s[Y,X]$ equal to the image of $[Y,X_s]\to[Y,X]$ under the composite projection $X_s\to X$. We clearly obtain a decreasing filtration, $F^{s+1}\subset F^s$, with $F^0=[Y,X]$. We can define $F^\infty[Y,X]$ as the intersection $\bigcap_{s=0}^{\infty} F^s[Y,X]$.

Proposition 3

If F_0^* is defined in terms of a resolution and F^* is defined in terms of any complex whatsoever, then $F_0^s\subset F^s$ for every s.

This is immediate by Proposition 2: we have a map of the resolution into the other complex, and the result follows from the definition.

As an obvious corollary, all resolutions of X define the same filtration on $[Y,X]$. Assuming that X actually has a resolution, we thus obtain a *canonical* filtration on $[Y,X]$.

Proposition 4

Every X admits an N-resolution for each N.

Here we assume that X is $(n-1)$-connected; the N-resolution we obtain will likewise be $(n-1)$-connected. To assure that $r_s<\infty$ for every s, we have assumed that $\pi_i(X)$ is a finitely generated group for every i. We will show how to construct the first stage of the resolution in such a way that X_1 has all the properties required of X; this will indicate how to build the desired resolution inductively.

Let $\{x_{i,j}\}$ be a set of generators for $H^i(X)$, $i\le N$, as an $\mathcal{A}$-module. Since all $\pi_i(X)$ are finitely generated, so are the $H^i(X)$, by the Hurewicz theorem mod $\mathcal{C}_{FG}$. We form the complex $\prod_{i,j}K(Z_2,n_{i,j})$, where $n_{i,j}$ is the dimension of $x_{i,j}$, and map X into this complex by a map g_0 such that $g_0^*(\iota_{i,j})=x_{i,j}$. Let X_1 be the induced fibre space over $X_0=X$ by g_0. From the homotopy exact sequence, it is clear that X_1 is $(n-1)$-connected and that its homotopy groups are finitely generated. The proposition now follows by induction.

Henceforth, the symbol $F^s[Y,X]$ will always be understood to refer to this canonical minimal filtration of $[Y,X]$.

We remark that the above construction is not designed to guarantee that the higher stages in the resolution will have higher connectivity than X. In fact, our resolution of $K(Z,n)$ will simply have $X_s=K(Z,n)$ for every s. It gives the filtration of $[S^n,K(Z,n)]=Z$ by powers of 2, that is $F^r[S^n,K(Z,n)]$ consists of the classes divisible by 2^r. But note that $F^\infty=0$ in this case.

EVALUATION OF F^∞

We now turn to the general problem of the evaluation of $F^\infty[Y,X]$. Under the usual assumptions on Y and X, $[Y,X]$ is an abelian group.

Therefore we can speak of a map $Y \to X$ as being "divisible by n" where n is any non-zero integer; this means that there exists $\alpha \in [Y,X]$ such that $n\alpha$ is the homotopy class of the given map.

Theorem 1

A class $\alpha \in [Y,X]$ lies in $F^{\infty}[Y,X]$ if and only if α is divisible by 2^r for every positive integer r.

We will prove this theorem in two parts: first, if α is divisible by 2^s, then $\alpha \in F^s[Y,X]$; second, if α is not divisible by 2^r, then there is some t for which α does not lie in $F^t[Y,X]$.

The first part is easy; the essential point is that $2[Y,K_s]=0$. We argue by induction on s, assuming the statement valid for F^{s-1} in all complexes over any nice space. (The assertion is trivial when $s=0$.) If $\alpha = 2^s\beta$, consider $2\beta \in [Y,X]$; this lies in the kernel of $(g_0)_{\#}: [Y,X] \to [Y,K_0]$, and therefore there exists a lifting $u: Y \to X_1$ with $p_1 u = 2\beta$. (As usual we use the same letter for a map and for its homotopy class.) Now

$$\alpha = 2^s\beta = 2^{s-1}(2\beta) = 2^{s-1}(p_1 u) = p_1(2^{s-1}u)$$

By the induction hypothesis, $2^{s-1}u$ lies in $F^{s-1}[Y,X_1]$, so that this map of Y into X_1 factors through X_s, which is to say that $\alpha \in F^s[Y,X]$, as was to be proved.

The second part of the proof requires the construction of a certain complex. Let ι denote the class of the identity map $X \to X$. Whether or not $[X,X]$ is a group depends on the space X, and we wish to consider the class $2^r(\iota)$. Since we work in the stable range, we can validate this without much trouble, say by assuming X to be $(n-1)$-connected and then considering $[X^{\cdot},X]$ instead of $[X,X]$, where $X^{\cdot}$ is an appropriate skeleton of X, for example, the $(2n-3)$-skeleton; ι then denotes the class of the inclusion. The reader should be able to handle the details of such matters.

Let U be the fibre space induced over SX—or $S(X^{\cdot})$—by the mapping $S(2^r\iota)$. The fibre is ΩSX, which agrees with X in the stable range, and we consider the diagram

$$\begin{array}{ccccc} X \xrightarrow{2^r\iota} X \sim \Omega SX & \xrightarrow{i} & U & & \\ & & \downarrow{\scriptstyle p} & & \\ & & SX & \xrightarrow{S(2^r\iota)} & SX \end{array}$$

which gives rise to an exact sequence

$$[Y,X] \xrightarrow{2^r} [Y,X] \xrightarrow{i_*} [Y,U] \xrightarrow{p_*} [Y,SX] \xrightarrow{2^r} [Y,SX]$$

under a suitable restriction on the dimension of Y. Thus $[Y,U]$ is a group extension of ker (2^r) by coker (2^r), that is, the sequence

$$0 \to \text{coker } (2^r) \to [Y,U] \to \text{ker } (2^r) \to 0$$

is exact. Since $2^r(\text{coker } (2^r)) = 0$ and $2^r(\text{ker } (2^r)) = 0$, it follows that $2^{2r}[Y,U] = 0$. We use this fact to construct a complex over U, killing its homotopy groups. If X is $(n-1)$-connected, then so is U. Let π_0 be the first nontrivial homotopy group of U, occurring in dimension $n_0 \geq n$. Then π_0 is an abelian group and there is a canonical epimorphism $\pi_0 \to \pi_0 \otimes Z_2$. Let $h_0 \colon U \to K_0$ be the corresponding map under the equivalence

$$\text{Hom } (\pi_0, \pi_0 \otimes Z_2) \leftrightarrow H^{n_0}(U; \pi_0 \otimes Z_2) \leftrightarrow [U,K_0]$$

where $K_0 = K(\pi_0 \otimes Z_2, n_0)$. Let U_1 be the induced fibre space over U defined by h_0. From the homotopy exact sequence of this fibre space, it is clear that $\pi_{n_0}(U_1)$ is isomorphic to $2(\pi_{n_0}(U))$. But $2^{2r}(\pi_m(U))$ is zero for all m in the stable range. Thus we may proceed with the above construction, killing $\pi_{n_0}(U)$ in not more than $2r$ steps. We can then continue to build the complex over U, killing the next non-trivial homotopy group, and we can annihilate $\pi_*(U)$ up through dimension m in a finite number of steps. Thus if the space Y has finite dimension m, we reach a t such that $[Y,U_t] = 0$.

With this machinery, it is easy to complete the second part of the proof of the theorem. If $\alpha \in [Y,X]$ is not divisible by 2^r, we cover the map $i \colon X \to U$ by a map of a resolution of X into the above complex over U. Then α is certainly not in $F^t[Y,X]$ for any t such that $[Y,U_t] = 0$, since from the diagram below, it is clear that $\alpha \in F^t[Y,X]$ implies $\alpha \in \text{ker } (i_*) = 2^r[Y,X]$. This completes the proof of Theorem 1.

$$\begin{array}{ccc} X_t & \xrightarrow{i_t} & U_t \\ \downarrow & & \downarrow \\ Y \xrightarrow{\alpha} X & \xrightarrow{i} & U \end{array}$$

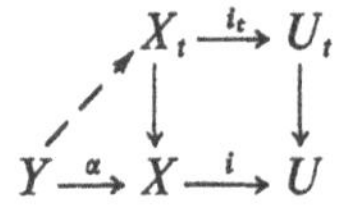

Suppose that $[Y,X]$ is a finitely generated abelian group; then it is expressible as a direct sum of a free abelian group and of finite cyclic groups of prime-power order. Clearly the elements infinitely divisible by 2 are just the elements of the odd-primary part of the group.

Our hypotheses assure that $[Y,X_s]$ will be a finitely generated abelian group for every $s \geq 0$. Indeed, Y is a finite complex with dim $Y \leq 2n - 2$ and each X_s is $(n-1)$-connected with finitely generated homotopy groups. Then the Postnikov spectral sequence of Chapter 14 converges to $[Y,X_s]$ and has only finitely generated groups in its initial term.

Under these hypotheses, then, $F^\infty[Y,X]$ is the direct summand of $[Y,X]$

consisting of torsion elements of odd order; in particular, it has no element of order 2. Further, $[Y,X]/F^{\infty}[Y,X]$ is the 2-component of $[Y,X]$.

THE STABLE GROUPS $\{Y,X\}$

For Y in the stable range with respect to X, we now have defined a filtration on $[Y,X]$ by means of a resolution of X. It would be tempting now to consider, in the manner of Chapter 14, an exact couple with typical D-term of form $[S^pY,X_q]$ and typical E-term $[S^pY,K_q]$, with suitable bigradings. Indeed, basically this is what we are going to do in order to construct the Adams spectral sequence.

However, some more preliminaries are still necessary. First of all, we need an algebraic detour so that we will recognize E_2 for what it is. We take this detour in the next section.

More importantly, the above candidates for D- and E-terms of our proposed spectral sequence take us back out of the stable range. Dim $Y \leq 2n-2$ does not imply dim $S^pY \leq 2n-2$.

However, if we replace n by $n+m$, for m sufficiently large with respect to p, dim $Y \leq 2n-2$ does imply dim $S^{p+m}Y \leq 2(n+m)-2$. This observation suggests replacing the $(n-1)$-connected X with the $(n+m-1)$-connected S^mX and Y with S^mY. We then would be studying a term in a spectral sequence for $[S^mY,S^mX]$, but since dim $Y \leq 2n-2$, this group is independent of m anyway.

We thus are led to consider a spectral sequence with typical D-term of form $[S^{p+m}Y,{}^mX_q]$ and typical E-term $[S^{p+m}Y,{}^mK_q]$, where ${}^mX_q$ and ${}^mK_q$ are the appropriate terms of a resolution of S^mX and where m is sufficiently large, depending on p and q. Of course, to justify this procedure we must show that the stable portions of the different exact couples obtained for different m agree under suspension.

Granting this (see Proposition 5), we now notice that the assumption dim $Y \leq 2n-2$ is no longer essential. Assuming only that Y is a finite complex, the groups $[S^mY,S^mX]$ are independent of m for m sufficiently large. The common value is denoted $\{Y,X\}$ and is called the (abelian) group of S-maps of Y into X.

We now state the result which shows independence of the choice of large m.

Proposition 5

Let $\mathfrak{X}$ be an $(n-1)$-connected resolution of X. Then there exists an n-connected resolution ${}^1\mathfrak{X}$ of SX such that SX_s and ${}^1X_s$ have the same $(2n-s)$-type for each s.

Thus for Y such that $\dim Y \leq 2n-2-s$, we have $[Y,X_s] = [SY,SX_s] = [SY,{}^1X_s]$. Hence for m sufficiently large, depending on p and q, $[S^{p+m}Y,{}^mX_q]$ is independent of m.

The proof of Proposition 5 is by induction on s. The assertion is trivial for $s=0$. We sketch the induction step, supposing ${}^1X_s$ has been constructed of the same $(2n-s)$-type as SX_s. Thus we may choose ${}^1K_s$ so that $\Omega({}^1K_s) = K_s$ through dimension $2n-s-2$; that is, for $i \leq 2n-s-2$, $\pi_i K_s = \pi_{i+1}{}^1K_s$, with the image of ${}^1g_s^{i+1}$ corresponding under suspension to the image of g_s^i. For such a choice of ${}^1g_s$ and ${}^1K_s$, calculation verifies that ${}^1X_{s+1}$ satisfies the induction hypothesis.

Of course the notations ${}^1X$, ${}^1g$, etc., have nothing to do with the cohomology suspension of Chapter 14.

THE DEFINITION OF EXT$_{\mathcal{A}}$ (M,N)

In this section we take the necessary algebraic detour so that we will recognize E_2 of our spectral sequence. Our goal is to define the symbol $\mathrm{Ext}_{\mathcal{A}}(M,N)$, where M and N are graded modules over the Steenrod algebra $\mathcal{A}$.

This symbol is defined by a standard construction of homological algebra. We will now sketch the pertinent algebraic theory. The main theorem of this chapter, on the existence and fundamental properties of the Adams spectral sequence, will tie this algebraic material to the above material on complexes, resolutions, and filtrations.

Let R be a graded ring with unit. Recall from Chapter 6 that a graded left R-module M is a sequence of unitary left R_0-modules M_i together with an associative action $\mu: R \otimes M \to M$. We will suppose $M_i = 0$ for $i < 0$. A *homomorphism of degree t* between two graded left R-modules is a sequence of left R_0-homomorphisms $f_q: M_q \to N_{q+t}$ respecting the actions μ_M and μ_N. Such a homomorphism may be abbreviated $f: M \to N$. We denote by $\mathrm{Hom}_R^t(M,N)$ the group of homomorphisms $M \to N$ of degree $-t$. A sequence

$$0 \leftarrow M \xleftarrow{\varepsilon} C_0 \xleftarrow{d_1} C_1 \xleftarrow{d_2} C_2 \leftarrow \cdots \xleftarrow{d_s} C_s \leftarrow \cdots$$

of graded left R-modules and homomorphisms is called a *projective resolution of M* if the following conditions are satisfied:

1. Each $C_s = \{C_{s,t}\}$ is a projective graded left R-module
2. All the homomorphisms $\{d_s\}$, ε are homomorphisms of degree zero
3. The sequence is exact

It is a standard theorem of homological algebra that such projective resolutions exist and are unique up to chain homotopy.

If $\{C_s\}$ is a projective resolution of M, and N is a graded left R-module, then in the sequence

$$0 \longrightarrow \mathrm{Hom}_R^t(C_0,N) \xrightarrow{d_1^*} \mathrm{Hom}_R^t(C_1,N) \longrightarrow \cdots \xrightarrow{d_s^*} \mathrm{Hom}_R^t(C_s,N) \cdots$$

we have $d_{s+1}^* d_s^* = 0$. The homology of this sequence at $\mathrm{Hom}_R^t(C_s,N)$ is denoted $\mathrm{Ext}_R^{s,t}(M,N)$; the groups $\mathrm{Ext}_R^{s,t}(M,N)$ form a bigraded group. Later we shall make some observations about how to calculate such things. For the moment we content ourselves with the remark that $\mathrm{Ext}_R^{0,t}(M,N) \approx \mathrm{Hom}_R^t(M,N)$, since the Hom functor applied to the exact sequence

$$0 \longleftarrow M \xleftarrow{\varepsilon} C_0 \xleftarrow{d_1} C_1$$

gives an exact sequence

$$0 \longrightarrow \mathrm{Hom}_R^t(M,N) \xrightarrow{\varepsilon^*} \mathrm{Hom}_R^t(C_0,N) \xrightarrow{d_1^*} \mathrm{Hom}_R^t(C_1,N)$$

for each t.

Under certain conditions, Ext as defined above can be given a multiplicative structure so that $\mathrm{Ext}^{s,t}$ becomes a bigraded algebra. In particular, if R is an augmented algebra over a field K, then $\mathrm{Ext}^{s,t}(K,K)$ becomes a bigraded algebra over K. (Here K is made a graded R-module, concentrated in gradation 0, by means of the augmentation.) We will be interested in the case $K = Z_2$, $R = \mathcal{A} =$ the mod 2 Steenrod algebra. In this case, R is a co-commutative Hopf algebra and the algebra $\mathrm{Ext}_{\mathcal{A}}(Z_2,Z_2)$ turns out to be commutative. We will not define the multiplication in Ext.

PROPERTIES OF THE SPECTRAL SEQUENCE

We are now in a position to formulate the main theorem on the Adams spectral sequence.

Theorem 2

Let X be a space such that $\pi_i(X)$ is finitely generated for every $i \geq 0$, and let Y be a finite complex. Then there exists a decreasing filtration F^* of $\{Y,X\}$ and a spectral sequence $\{E_r,d_r\}$ with the following properties:

1. Each E_r is a bigraded group, and d_r is a homomorphism of bigraded groups, $d_r^{s,t}: E_r^{s,t} \longrightarrow E_r^{s+r,t+r-1}$ such that $d_r d_r = 0$
2. E_2 is naturally isomorphic to $\mathrm{Ext}_{\mathcal{A}}(H^*(X),H^*(Y))$ (cohomology with coefficients in Z_2 as usual)

3. There is a natural monomorphism $E_{r+1}^{s,t} \to E_r^{s,t}$ whenever $r > s$, and $E_\infty^{s,t} = \bigcap_{r>s} E_r^{s,t}$ (where E_∞ will be defined in the proof of this assertion below)
4. $E_\infty^{s,t}$ is naturally isomorphic to $F^s\{S^{t-s}Y,X\}/F^{s+1}\{S^{t-s}Y,X\}$
5. $F^\infty\{Y,X\}$ consists of the subgroup of elements of finite odd order

Some parts of this theorem have essentially been proved already. Before completing the proof of properties (1) to (5), we state here for completeness some further results for the case $X = S^0$, $Y = S^0$, which pertains, by (4) above, to the stable homotopy of spheres. In this case, by (2) above, $E_2 = \mathrm{Ext}_{\mathcal{A}}(Z_2,Z_2)$.

6. Every E_r is endowed with a multiplicative structure compatible with the bigrading, and d_r is a derivation
7. The product in E_2 is the usual product in Ext (which is associative and commutative)
8. The product in E_{r+1} is induced from that in E_r by passing to quotients (and hence is also associative and commutative)
9. The product in E_∞ may equally well be regarded as induced from that in E_r, using (3) and (8), or as obtained by passing to quotients in the composition ring of stable homotopy of spheres, using (4)

This theorem has become one of the cornerstones of stable homotopy theory. It separates the problem of calculating the stable homotopy groups of spheres, for example, into three distinct steps: the purely algebraic problem of calculating $\mathrm{Ext}_{\mathcal{A}}(Z_2,Z_2)$; the determination of the differentials in the spectral sequence; and the reconstruction of the group extensions and products which are lost in passing to quotients—as in (4).

We now proceed with the construction of the spectral sequence and the proof of properties (1) to (5).

The spaces X and Y are not assumed to have any connectedness properties, since we are interested only in stable calculations. For the most part, we will be working in the proof with S^nY and S^nX, instead of Y and X. Then of course we are dealing with spaces which are $(n-1)$-connected.

We begin by taking a resolution of S^nX, in which we denote the stages of the resolution by ${}^nX_s$ $(s \geq 0)$ and the corresponding generalized Eilenberg-MacLane spaces by ${}^nK_s$ $(s \geq 0)$. Recall that by Proposition 5 a resolution of $S^{n+m}X$ for the stable range can be assumed to look like the mth suspension of our resolution of S^nX.

For any non-negative m we obtain a filtration on the homotopy classes $[S^{n+m}Y,S^nX]$ in the manner which has been discussed earlier. Together with our previous remarks about the stability of the resolution, this defines the stable filtration of the theorem and proves (5).

Let Y' denote an appropriate suspension of Y. Then the homotopy classes of maps of Y into the various spaces in the resolution of S^nX (and their loop spaces) form a commutative diagram

$$\begin{array}{cccccccccc}
 & & & \vdots & & & & \vdots & & \\
 & & & \downarrow & & & & \downarrow & & \\
\cdots \to [S^2Y', {}^nK_0] & \to & & [SY', {}^nX_1] & \to & [SY', {}^nK_1] & \to & [Y', {}^nX_2] & \to & [Y', {}^nK_2] \\
 & & & \downarrow & & & & \downarrow & & \\
 & & & [SY', S^nX] & \to & [SY', {}^nK_0] & \to & [Y', {}^nX_1] & \to & [Y', {}^nK_1] \\
 & & & & & & & \downarrow & & \\
 & & & & & & & [Y', S^nX] & \to & [Y', {}^nK_0]
\end{array}$$

as in Chapter 14. From this diagram we can form a bigraded exact couple with

$$D_1^{s,t} = [S^{n+t-s}Y, {}^nX_s]$$
$$E_1^{s,t} = [S^{n+t-s}Y, {}^nK_s]$$

(with the convention that ${}^nX_0 = S^nX$). The maps i,j,k of the exact couple, in the notation of Chapter 7, are induced by the projections, the "k-invariants," and the inclusions of the fibres. The map i has bidegree $(-1,-1)$; the map j has bidegree $(0,0)$; the map k, and hence also the first differential $d = jk$, has bidegree $(1,0)$. (We remark that the bigrading used here is not the same as that used in Chapter 14.)

We have defined $E_1^{s,t}$ only for non-negative s, but we can put $E_1^{s,t} = 0$ for $s < 0$. This is tantamount to continuing the resolution below S^nX by a string of one-point spaces.

The rth differential of the exact couple may be roughly described as $j \circ i^{-r+1} \circ k$, and it obviously has bigrading $(r, r-1)$. We have thus verified (1).

It remains to prove (2), (3), and (4). We will now undertake the proof of (2).

Lemma 1

$[S^{n+t-s}Y, {}^nK_s]$ is naturally isomorphic to $\mathrm{Hom}_{\mathcal{A}}^{t-s}(H^*({}^nK_s), H^*(S^nY))$.

Every map between two spaces induces a map on the cohomology level which is an $\mathcal{A}$-module homomorphism of degree zero. In this way we obtain a transformation $[S^mY, {}^nK_s] \to \mathrm{Hom}_{\mathcal{A}}^0(H^*({}^nK_s), H^*(S^mY))$. We have proved in Chapter 1 that this is an isomorphism in the case when ${}^nK_s$ is a $K(\pi,n)$ space. It is not hard to establish this isomorphism in the present case.

We omit the details. The result then follows from the observation that

$$\mathrm{Hom}^0_{\mathcal{A}}\,(H^*({}^nK_s),H^*(S^mY)) \approx \mathrm{Hom}^k_{\mathcal{A}}\,(H^*({}^nK_s),H^*(S^{m-k}Y))$$

(using the suspension isomorphism) by putting $m=n+t-s$ and $k=t-s$.

Thus $d_1 : E_1^{s,t} \to E_1^{s+1,t}$ may be considered as a map

$$\mathrm{Hom}^{t-s}_{\mathcal{A}}\,(H^*({}^nK_s),H^*(S^nY)) \to \mathrm{Hom}^{t-s-1}_{\mathcal{A}}\,(H^*({}^nK_{s+1}),H^*(S^nY))$$

and it is clear from the definition of d_1 and the proof of Lemma 1 that in this interpretation d_1 is simply the map induced by going from ΩK_s to K_{s+1} in the (expanded) resolution.

Now $\mathrm{Ext}^{s,t}_{\mathcal{A}}\,(H^*X,H^*Y) = \mathrm{Ext}^{s,t}_{\mathcal{A}}\,(H^*(S^nX),H^*(S^nY))$ by the suspension isomorphism, and by definition this latter group is the homology of the sequence $\{\mathrm{Hom}^t_{\mathcal{A}}\,(C_s,H^*(S^nY))\}$ where the $\{C_s\}$ form a projective resolution of $H^*(S^nX)$ in the algebraic sense. We claim that such a resolution is obtained from $\{H^*({}^nK_s)\}$ simply by modifying the gradation. In fact, the expanded resolution over S^nX gives, on the cohomology level, a diagram (below) with each p^* trivial, each g^* epimorphic, and hence each i^* mono-

$$\begin{array}{ccccccccc} & & \uparrow{\scriptstyle p^*} & & & & \uparrow{\scriptstyle p^*} & & \\ H^*(\Omega^2({}^nK_s)) & \xleftarrow{i^*} & (\ \ ) & \xleftarrow{g^*} & H^*(\Omega({}^nK_{s+1})) & \xleftarrow{i^*} & (\ \ ) & \xleftarrow{g^*} & H^*({}^nK_{s+2}) \\ & & \uparrow{\scriptstyle p^*} & & & & \uparrow{\scriptstyle p^*} & & \end{array}$$

morphic. The i–g–p sequences are exact, coming from the mapping sequences. It is easy to verify from this that the sequence

$$H^*(\Omega^2({}^nK_s)) \leftarrow H^*(\Omega({}^nK_{s+1})) \leftarrow H^*({}^nK_{s+2})$$

is exact. But in the stable range this is exactly

$$H^{*-2}({}^nK_s) \xleftarrow{d_1} H^{*-1}({}^nK_{s+1}) \xleftarrow{d_1} H^*({}^nK_{s+2})$$

with the reinterpretation of d_1 given above. In this way we find that

$$0 \leftarrow H^*(S^nX) \leftarrow H^*({}^nK_0) \xleftarrow{d_1} H^*({}^nK_1) \xleftarrow{d_1} H^*({}^nK_2) \leftarrow \cdots$$

is a projective resolution (in the stable range) of $H^*(S^nX)$, except that the homomorphisms all have degree 1 instead of degree zero as required for a resolution. We thus obtain a resolution if we replace each $H^*({}^nK_s)$ by a module C_s isomorphic to it but graded so that each element of C_s has gradation less by s than the corresponding element of $H^*({}^nK_s)$. We then have that $\mathrm{Ext}^{s,t}_{\mathcal{A}}\,(H^*X,H^*Y)$ is the homology of the sequence

$$\{\mathrm{Hom}^t_{\mathcal{A}}(C_s,H^*(S^nY))\},$$

but this is the same as the sequence $\{\mathrm{Hom}^{t-s}_{\mathcal{A}}\,(H^*({}^nK_s),H^*(S^nY))\}$, and we have seen that the homology of this sequence is just E_2.

This proves (2). Observe, as a corollary, that E_2 depends on X and Y only through H^*X and H^*Y as $\mathcal{A}$-modules.

In order to prove (3), we will follow much the same route as was taken in Chapter 7 in the discussion of the homology exact couple of a filtered complex.

Lemma 2

$E_r^{s,t}$ is isomorphic to the quotient

$$\frac{A}{B} = \frac{\text{im}: [Y', {}^nX_s/{}^nX_{s+r}] \to [Y', {}^nX_s/{}^nX_{s+1}]}{\text{im}: [SY', {}^nX_{s-r+1}/{}^nX_s] \to [Y', {}^nX_s/{}^nX_{s+1}]}$$

where the map in the numerator A is induced by the natural map ${}^nX_s/{}^nX_{s+r} \to {}^nX_s/{}^nX_{s+1}$ (each projection $X_{s+1} \to X_s$ being interpreted as an inclusion), and the map in the denominator B is from the exact fibre mapping sequence obtained by mapping Y' into successive pairs from the triple $({}^nX_{s-r+1}, {}^nX_s, {}^nX_{s+1})$; here Y' denotes $S^{n+t-s}Y$.

We will not give the details of the proof of this lemma because it is formally the same as the proof of the corresponding result in Chapter 7 (Proposition 3), with the appropriate modifications. The essential observations are the exactness of the sequence arising from the triple, since we are in the stable range, and the identification of ${}^nX_s/{}^nX_{s+1}$ with ${}^nK_s$ in the stable range.

We now define E_∞ as follows.

Definition

$$E_\infty^{s,t} = \frac{\text{im}: [Y', {}^nX_s] \to [Y', {}^nK_s]}{\text{im}: [SY', S^nX/{}^nX_s] \to [Y', {}^nK_s]}$$

with the notations and conventions described above. The map in the denominator comes from the triple $(S^nX, {}^nX_s, {}^nX_{s+1})$.

This definition is motivated by the formal relationship with E_r as characterized in Lemma 2 and by part (3) of the theorem, the proof of which is now well under way.

Although an n occurs in this definition of E_∞ (as indeed in all the definitions which we have given), it does not play an essential role, since we confine our interest to the stable range, in which the results are independent of n.

We can now complete the proof of (3). Recall that $E_1^{s,t} = 0$ if $s < 0$, and thus the same is true for all E_r. Since d_r has bidegree $(r, r-1)$, no differentials map into $E_r^{s,t}$ if $r > s$. Therefore there is a canonical monomorphism $E_{r+1}^{s,t} \to E_r^{s,t}$ whenever $r > s$, and the symbol

$$\bigcap_{r>s} E_r^{s,t}$$

has a meaning.

We denote by

$$\alpha_s\colon F^\infty[Y',{}^nX_s] \to F^\infty[Y',{}^nX_{s-1}]$$

the restriction of $(p_s)_\#$ where $p_s\colon {}^nX_s \to {}^nX_{s-1}$.

Lemma 3

Every α_s is monomorphic.

This follows from the exact sequence

$$[Y',\Omega({}^nK_{s-1})] \xrightarrow{i_\#} [Y',{}^nX_s] \xrightarrow{(p_s)_\#} [Y',{}^nX_{s-1}]$$

because, since $[Y',{}^nX_s]$ is a finitely generated abelian group, nothing in $F^\infty[Y',{}^nX_s]$ is of order 2, whereas everything in $\text{im}\,(i_\#) = \ker\,((p_s)_\#)$ is of order 2. Thus the intersection of the domain of α_s with the kernel of $(p_s)_\#$ is zero, which proves the lemma.

Part (3) of the theorem will follow from the next lemma.

Lemma 4

In the notation of the diagram

$$\begin{array}{ccccccc}
 & & [Y',{}^nX_s/{}^nX_{s+r}] & \longrightarrow & [Y',S({}^nX_{s+r})] & & \\
 & \nearrow & \downarrow f_r & & \downarrow g_r & \searrow & \\
[Y',{}^nX_s] & & & & & & [Y',S({}^nX_s)] \\
 & \searrow f_\infty & & & & \nearrow p_\# & \\
 & & [Y',{}^nX_s/{}^nX_{s+1}] & \xrightarrow{\partial} & [Y',S({}^nX_{s+1})] & &
\end{array}$$

and with the notations and conventions which have gone before, the following statements are equivalent:

i. $\text{im}\,(f_\infty) = \bigcap \text{im}\,(f_r)$
ii. $\text{im}\,(f_\infty) = \partial^{-1}(\bigcap \text{im}\,(g_r))$
iii. $\partial^{-1}(0) = \partial^{-1}(\bigcap \text{im}\,(g_r))$
iv. $0 = \text{im}\,(\partial) \cap (\bigcap \text{im}\,(g_r))$
v. $0 = \ker\,(p_\#) \cap F^\infty[Y',S({}^nX_{s+1})]$
vi. Each α_s is monomorphic

PROOF: We leave to the reader the verification that the diagram is commutative and that its rows are exact. The equivalence of (ii) with (iii) and of (iv) with (v) follow from exactness and from the definition of the filtration. The equivalence of (iii) with (iv) is elementary set theory. The equivalence of (v) with (vi) follows from the definition of α_s. It remains to prove the equivalence of (i) with (ii). Now $\operatorname{im}(f_r) = \partial^{-1}(\operatorname{im}(g_r))$ by a simple "diagram-chase." Therefore

$$\bigcap \operatorname{im}(f_r) = \bigcap \partial^{-1}(\operatorname{im}(g_r)) = \partial^{-1}\left(\bigcap \operatorname{im}(g_r)\right)$$

and the equivalence of (i) with (ii) follows.

This proves the lemma.

Now we have already proved (vi), in Lemma 3. Thus (i) has been established. But (i) implies part (3) of the theorem. For, if we denote by B_r the denominator of the quotient in Lemma 2, we have, by that lemma,

$$E_r = \frac{\operatorname{im}(f_r)}{B_r}$$

whereas, by definition,

$$E_\infty = \frac{\operatorname{im}(f_\infty)}{B_\infty}$$

with $B_\infty = \lim_r(B_r)$ under an obvious convention. (B_r becomes constant when $r > s$.) From (i) of the last lemma, part (3) of the theorem follows.

It remains to prove (4). From the diagram below and a standard argu-

$$\begin{array}{ccccc} [Y',{}^nX_{s+1}] & & & & [SY',S^nX/{}^nX_s] \\ \downarrow & \searrow & & \swarrow & \downarrow \\ & & [Y',{}^nX_s] & & \\ & \swarrow & & \searrow & \\ [Y',S^nX] & & & & [Y',{}^nK_s] \end{array}$$

ment, already used in Chapter 7 (see Exercise 1 of that chapter), we have the isomorphism

$$E_\infty^{s,t} \approx \frac{\operatorname{im}\colon [Y',{}^nX_s] \to [Y',S^nX]}{\operatorname{im}\colon [Y',{}^nX_{s+1}] \to [Y',S^nX]}$$

which proves (4).

This completes the proof of (1) to (5) and establishes the fundamental properties of the Adams spectral sequence.

MINIMAL RESOLUTIONS

When the algebra $\mathcal{A}$ is as complicated as the Steenrod algebra, $\operatorname{Ext}_{\mathcal{A}}(H^*X, H^*Y)$ may be very complicated and may present a considerable task in computation. In fact, the computation of $\operatorname{Ext}_{\mathcal{A}}(Z_2, Z_2)$ has been the subject of much recent research.

We will make some observations on the calculation of $\mathrm{Ext}_{\mathcal{A}}(M,Z_2)$, where M is a left $\mathcal{A}$-module, before specializing to the case $M=Z_2$. Although we are using $\mathcal{A}$ to denote the mod 2 Steenrod algebra, the following remarks are valid if $\mathcal{A}$ is any connected algebra over Z_2 (and in fact can be generalized much further).

The augmentation of $\mathcal{A}$ is a map $\varepsilon\colon \mathcal{A}\to Z_2$, of which the kernel contains all of $\mathcal{A}$ except the multiplicative unit. This kernel is called the "augmentation ideal" and is denoted $I(\mathcal{A})$.

If M is a left $\mathcal{A}$-module, we write $J(M)$ for $I(\mathcal{A})M$, that is, for the set of all finite linear combinations $\sum a_i m_i$ where $a_i \in I(\mathcal{A})$ and $m_i \in M$. This is a sub-$\mathcal{A}$-module of M.

Definition

An $\mathcal{A}$-module homomorphism $f\colon M\to N$ is *minimal* if ker (f) is contained in $J(M)$.

A projective resolution is called a *minimal resolution* if all the homomorphisms in the resolution are minimal.

Proposition 6

Every graded left $\mathcal{A}$-module M admits a minimal resolution.

It is enough to construct one stage, i.e., to find a free graded left $\mathcal{A}$-module C and a minimal epimorphism $d\colon C\to M$. (The proof is then completed by induction.) Let $\{g_i\}$ be a minimal set of generators for M as an $\mathcal{A}$-module. Then let C be a free $\mathcal{A}$-module on the symbols $\{h_i\}$ where each h_i has the same grading as the corresponding g_i. There is an $\mathcal{A}$-module homomorphism of degree zero, $d\colon C\to M$, which takes h_i to g_i for each i. Since $\{g_i\}$ is a generating set for M as an $\mathcal{A}$-module, d is obviously an epimorphism. To see that d is minimal, suppose that the kernel of d contains an element $b=\sum a_i h_i$ not in $J(M)$; then we may suppose that a_1 is not in $I(\mathcal{A})$, that is, that a_1 is the unit of $\mathcal{A}$. Then $0=d(b)=d(\sum a_i h_i)=\sum a_i g_i$, but this implies that $g_1=\sum_{i\neq 1} a_i g_i$, which contradicts the choice of $\{g_i\}$ as a minimal set of generators.

Proposition 7

Let

$$0\leftarrow M\xleftarrow{\varepsilon} C_0\xleftarrow{d_1} C_1\leftarrow\cdots\xleftarrow{d_s} C_s\leftarrow\cdots$$

be a minimal resolution of M. Then the homomorphisms

$$d_s^*\colon \mathrm{Hom}^t_{\mathcal{A}}(C_{s-1},Z_2)\to \mathrm{Hom}^t_{\mathcal{A}}(C_s,Z_2)$$

are all zero homomorphisms (for every $s\geq 1$).

PROOF: First observe that, by the definition of Z_2 as a left $\mathcal{A}$-module, $I(\mathcal{A})Z_2=0$.

If $g \in \mathrm{Hom}^t_{\mathcal{A}}(C_{s-1},Z_2)$, then $d_s^*(g)$ is defined by the equation

$$d_s^*(g)(c)=g(d_s(c)) \qquad c \in C_s$$

and we wish to show that this vanishes for all g and c. Now $d_{s-1}d_s=0$ for every $s \geq 2$ (and $\varepsilon d_1=0$), so that $d_s(c) \in \ker(d_{s-1})$ (and $d_1(c) \in \ker(\varepsilon)$). By the minimality hypothesis, this implies $d_s(c) \in J(C_{s-1})$ (and $d_1(c) \in J(M)$). But g is an $\mathcal{A}$-module homomorphism, and so g must annihilate $J(C_{s-1})$ (and $J(M)$), by the opening remark of the proof. This shows that indeed $d_s^*(g)(c)=0$ for every $c \in C_s$ $(s \geq 1)$, so that $d_s^*(g)=0$, but this holds for all g, so that $d_s^*=0$ (for every $s \geq 1$). Thus the proposition is proved.

The usefulness of minimal resolutions is now apparent from the following consequence.

Corollary 1

If the $\{C_s\}$ form a minimal resolution of M, then $\mathrm{Ext}^{s,t}_{\mathcal{A}}(M,Z_2)=\mathrm{Hom}^t_{\mathcal{A}}(C_s,Z_2)$.

This is immediate from the definition of Ext and from the last proposition.

SOME VALUES OF EXT$^{s,t}_{\mathcal{A}}$ (Z_2,Z_2)

In order to see how the machinery works, we look at the Steenrod algebra $\mathcal{A}$ and take $M=Z_2$. This is the E_2 term of the Adams spectral sequence for $X=Y=S^0$ and is therefore related to the stable homotopy of spheres.

To begin the calculation of $\mathrm{Ext}_{\mathcal{A}}(Z_2,Z_2)$, we observe that C_0 in a minimal resolution of Z_2 over $\mathcal{A}$ should be a free $\mathcal{A}$-module on one generator, so that $C_0 \approx \mathcal{A}$ and $\varepsilon: C_0 \to Z_2$ is just the augmentation of $\mathcal{A}$. Thus $\mathrm{Ext}^{0,t}_{\mathcal{A}}(Z_2,Z_2)$ vanishes for $t \neq 0$, while $\mathrm{Ext}^{0,0}_{\mathcal{A}}(Z_2,Z_2)=Z_2$, generated by the unit 1 of $\mathrm{Ext}_{\mathcal{A}}(Z_2,Z_2)$.

The next term, C_1, should be a free $\mathcal{A}$-module with one generator for each element of a minimal set of generators of the $\mathcal{A}$-module $I(\mathcal{A})$. Such a minimal set is given by $\{Sq^{2^i}\}$ $(i \geq 0)$, since these elements generate $\mathcal{A}$ as an algebra and are indecomposable in $\mathcal{A}$. Thus C_1 should be a free $\mathcal{A}$-module with one generator in gradation $t=2^i$ for each $i \geq 0$. Each such generator gives a homomorphism of graded left $\mathcal{A}$-modules of degree $(-t)$ from C_1 to Z_2, and thus $\mathrm{Ext}^{1,t}_{\mathcal{A}}(Z_2,Z_2)=\mathrm{Hom}^t_{\mathcal{A}}(C_1,Z_2)$ has a generator for each t which is a power of 2. The usual notation for these elements in Ext is h_i.

9	h_0^9																	P^2h_1
8	h_0^8															$h_0{}^7h_4$		Ph_1c_0
7	h_0^7											$Ph_0{}^2h_2$ $=Ph_1{}^3$				$h_0{}^6h_4$	Pc_0	$h_0{}^3e_0$ $=h_1{}^3d_0$
6	h_0^6										$Ph_1{}^2$	Ph_0h_2			$h_0{}^2d_0$	$h_0{}^5h_4$	$h_1{}^2d_0$ $=c_0{}^2$	$h_0{}^2e_0$
5	h_0^5									Ph_1		Ph_2			h_0d_0	$h_0{}^4h_4$ / h_1d_0		h_0e_0 $=h_2d_0$
4	h_0^4							$h_0{}^3h_3$		h_1c_0					d_0	$h_0{}^3h_4$		e_0
3	h_0^3			$h_0{}^2h_2$ $=h_1{}^3$				$h_0{}^2h_3$	c_0	$h_1{}^2h_3$ $=h_2{}^3$					$h_0h_3{}^2$	$h_0{}^2h_4$		$h_1{}^2h_4$
2	h_0^2		h_1^2	h_0h_2			h_2^2	h_0h_3	h_1h_3						h_3^2	h_0h_4	h_1h_4	
1	h_0	h_1		h_2				h_3								h_4		
$s=0$	1																	
$t-s=$	0	1	2	3	4	5	6	7	8	9	10	11	12	13	14	15	16	17

Figure 1 $\mathrm{Ext}_{\mathcal{A}}^{s,t}(Z_2,Z_2)$ for $t-s\leq 17$

It is considerably more difficult to write down the structure of C_2 and thus of $\mathrm{Ext}^{2,t}_{\mathscr{A}}(Z_2,Z_2)$. We will not go any further with the details but will be content with a statement of results.

Recall that $\mathrm{Ext}^{s,t}_{\mathscr{A}}(Z_2,Z_2)$ has a multiplicative structure under which it is an algebra over Z_2. The following results are proved by purely algebraic methods. In any finite range of dimensions, they can be verified by calculating in a minimal resolution, but this method has many limitations and has been superseded in practice by deeper algebraic methods.

Theorem 3

For $s=1$, $\mathrm{Ext}=\mathrm{Ext}^{s,t}_{\mathscr{A}}(Z_2,Z_2)$ has as Z_2-basis the elements $h_i \in \mathrm{Ext}^{1,t}$ with $t=2^i$. For $s=2$, Ext is generated by the products $h_i h_j$, subject only to the relations $h_i h_{i+1}=0$ $(i \geq 0)$. For $s=3$, the products $h_i h_j h_k$ are subject only to the additional relations $h_i h_{i+2}^2=0$ and $h_i^3=h_{i-1}^2 h_{i+1}$ and the relations implied by $h_i h_{i+1}=0$.

However, there are other generators for Ext with $s=3$, the first of which is a generator c_0 in bigrading $(s=3,\ t=11)$. A complete set of generators of $\mathrm{Ext}^{s,t}_{\mathscr{A}}(Z_2,Z_2)$ for $t-s \leq 17$ is presented in Figure 1.

In Figure 1 it is to be understood that there is a non-zero generator $h_0^s \in \mathrm{Ext}^{s,s}$ for every $s \geq 0$, but otherwise all generators in the range $t-s \leq 17$ are explicitly shown. (It is not hard to see that $\mathrm{Ext}^{s,t}=0$ if $t<s$.)

We might remark that to obtain this much of Ext by hand calculation of a minimal resolution requires several days.

APPLICATIONS TO HOMOTOPY GROUPS

We have arranged Figure 1 so that the bigradings of Ext for fixed $t-s$ appear in a column. This arrangement is motivated by the assertion of Theorem 2 that these are the groups which are associated to the stable $(t-s)$-stem of the homotopy of spheres. In fact, parts (4) and (5) of Theorem 2 assert that the groups $E_\infty^{s,t}$, with $t-s$ fixed, are the components of a normal series for the 2-primary component of the stable homotopy group π_{t-s}.

By part (1) of Theorem 2, d_r raises s by r while decreasing $(t-s)$ by 1. We will show now that the calculation of the differentials in the spectral sequence presents no difficulty whatsoever in the range $t-s \leq 14$. We will thus be able to write down a normal series for the 2-component of the first 13 stable homotopy groups of spheres, assuming Theorem 2 (parts (1) to (9)) and the calculation of Ext for the range of Figure 1.

The differential $d_r(h_1)$, if defined, must be h_0^{r+1} or zero, since it goes from $E_r^{1,2}$ to $E_r^{r+1,r+1}$. But the relation $h_0h_1=0$ implies that this differential is zero; for $d_r(h_0)=0$ for dimensional reasons (since $E_r=0$ when $t<s$ and d_r is a derivation (part (6) of the theorem), so that

$$0=d_r(0)=d_r(h_0h_1)=h_0(d_r(h_1))$$

whereas h_0^s is non-zero for all s. Thus $d_r(h_1)$ cannot be h_0^{r+1} and must be zero. Note also that h_1 cannot be in the image of any differential, since E_r vanishes for $s<0$.

This shows that the 2-component of the stable 0-stem is infinite. Moreover, it is possible to identify the image of h_0 in E_∞ with $2\iota: S^n \to S^n$, and on this basis it can be shown that multiplication by h_0 represents a non-trivial group extension in homotopy. Granting this, we deduce that the 2-component of the 0-stem is an infinite cyclic group Z. The 2-component of the stable 1-stem is clearly Z_2, generated by the image of h_1 in E_∞. We showed earlier that this generator is the (stable) Hopf map η.

Now dimensional considerations are enough to rule out any non-zero differentials originating in $3 \le t-s \le 7$. Thus the 2-stem is Z_2 generated by h_1^2, that is, by η^2 (cf. part (9) of the theorem), the 3-stem is Z_8 generated by the image of h_2 (which therefore corresponds to the Hopf map ν), and the 4-stem and 5-stem are empty. The relation $\eta^3=4\nu$, proved in Chapter 17, appears already in Ext as $h_1^3=h_0^2h_2$. The 6-stem is Z_2 generated by h_2^2, i.e., by ν^2.

Since $d_2(h_1h_3)$ lands in $E_2^{4,11}$, which is a non-trivial bigrading with a generator $h_0^3h_3$, dimensional considerations do not settle this differential. However, both $d_r(h_1)$ and $d_r(h_3)$ are zero for all $r \ge 2$, and since d_r is a derivation, this differential too is zero. Thus the 7-stem is Z_{16} generated by h_3, which corresponds to the Hopf map σ.

In the 8-stem there are two survivors to E_∞, namely, h_1h_3 and an element which we have denoted c_0, not expressible as a product of the h_i. It is perhaps not immediately clear whether the group extension here is trivial, i.e., whether the 8-stem should be Z_4 or Z_2+Z_2. However, the fact that c_0 and $h_0(h_1h_3)$ $(=0)$ have the same filtration degree $s=3$ implies that the extension is trivial and the 8-stem is Z_2+Z_2. For (9) asserts that the products in E_∞ are obtained from those in the homotopy composition ring by passing to quotients according to the degree s, and here, since there is no change in filtration, there can be no alteration of the product.

We caution the reader, however, not to assume too much about the relationship between the products in E_∞ and the homotopy products. For example, the fact that h_1h_3 survives the spectral sequence does not mean that the corresponding homotopy element is the composition product of the Hopf maps η and σ which are the images of h_1 and h_3. In fact, "the corresponding homotopy element" is not well-defined. What we can say about the 8-stem is this. There are three non-zero elements, each of order 2. One of these elements projects to zero in $E_\infty^{2,10}$; thus its projection to $E_\infty^{3,11}$ is defined and equal to c_0. The other two non-zero elements project to h_1h_3 in $E_\infty^{2,10}$. One of these elements is $\eta\sigma$; the other is $\eta\sigma$ plus the element which projects to c_0.

There obviously cannot be any non-zero differentials originating in the range $9 \leq t-s \leq 14$. We find that the 9-stem (or its 2-component, as always) is some group extension of three Z_2 groups, and the argument indicated above for the 8-stem applies here as well to show that the extensions are trivial. The 10-stem is Z_2 and the 11-stem is Z_8.

Following Adams, we have used the symbol P in the notation for the 9-stem and those following. There is a complicated and irregular periodicity in the stable homotopy of spheres, which is in evidence near the uppermost edge of the table of Ext, of which Figure 1 shows a portion. For our purposes we may simply regard P as an operator which goes from $E^{s,t}$ to $E^{s+4,t+12}$.

The 2-components of the 12- and 13-stems must be zero since there are no non-zero elements in E_2. However, after this point the work begins. There are several possible differentials between the 15-stem and the 14-stem, and indeed some of them turn out to be non-zero. (It is remarkable that the topology of spheres has waited so long before intruding into the algebraic picture.)

In the 14-stem we find that E_2 contains h_3^2 and also $h_0h_3^2$. If both these elements survived to E_∞, then it would follow that $\sigma^2: S^{n+14} \to S^{n+7} \to S^n$ was of order 4 (at least). However, the composition ring is in general anti-commutative, and since σ is in an odd stem (the 7-stem), it follows that $\sigma\sigma = -\sigma\sigma$ or $2\sigma^2 = 0$. Thus these two elements of E_2 cannot both survive to E_∞, and the only possibility is that $d_2(h_4) = h_0h_3^2$. In this way Adams obtained a new proof of Toda's theorem that there is no element of Hopf invariant one in the 15-stem, since elements of Hopf invariant one appear in filtration degree $s = 1$ and project to h_i.

There is also a non-zero d_3 at this point, namely, $d_3(h_0h_4) = h_0d_0$ (and of course $d_3(h_0^2h_4) = h_0^2d_0$), where d_0 is a new kind of generator of Ext appearing in $\mathrm{Ext}^{4,18}$. It can be shown that the only other non-zero differential originating in the portion of Ext shown in Figure 1 is $d_2(e_0) = h_1^2d_0$. Thus the 14-stem is a group extension of two Z_2 groups; the 15-stem is $Z_{32} + Z_2$ and the 16-stem is an extension of two Z_2 groups.

DISCUSSION

The above results give some idea of the workings of the machinery which we have set up and serve to suggest that the Adams spectral sequence can be a very useful weapon for the calculation of stable homotopy groups. As we write this, the complete description of the E_2 term in the above spectral sequence is not known, and no effective procedure is known for obtaining the differentials or reconstructing the group extensions, so that this technique is the subject of much current research.

This spectral sequence is of course due to Adams, who proved Theorem 2 on its existence and basic properties. We have not defined the multiplication in Ext (which is often called the cohomology of the Steenrod algebra and denoted $H^*(\mathcal{A})$), much less proved statements (6) to (9) of Theorem 2. The reader will find a proof in Adams' paper [1].

In this fundamental paper Adams also made the first important calculations of $H^*(\mathcal{A})$, namely, those of Theorem 3. They are not extensive, and further, very little was known at that time about the differentials.

Since then a great deal of progress has been made. Adams has proved various general structural theorems on $H^*(\mathcal{A})$, while May and others have introduced and refined new computational techniques, making possible extensive calculation in a relatively short time.

The differentials too have been subjected to attack. In particular, Adams, Maunder, May, and Mahowald have among them settled many cases, concerning both specific low-dimensional elements and certain periodic families of elements recurring in arbitrarily high stems.

REFERENCES

General properties of Ext

1. H. Cartan [6].
2. — and S. Eilenberg [1].
3. R. Godement [1].
4. S. MacLane [1].

Construction of the spectral sequence

1. J. F. Adams [1,4].
2. H. Cartan [6].

The cohomology of the Steenrod algebra

1. J. F. Adams [1,2,4].
2. A. Bousfield, *et al.* [1].
3. J. P. May [1].
4. M. C. Tangora [1].

Differentials in the Adams spectral sequence

1. J. F. Adams [2,4].
2. M. E. Mahowald [1].
3. — and M. C. Tangora [1].
4. C. R. F. Maunder [2].

BIBLIOGRAPHY

Adams, J. F.

1. *On the structure and applications of the Steenrod algebra*, Comm. Math. Helv. *32* (1958), 180–214.
2. *On the non-existence of elements of Hopf invariant one*, Ann. Math. (2) *72* (1960), 20–104.
3. *Vector fields on spheres*, Ann. Math. *75* (1962), 603–632.
4. *Stable homotopy theory*, Lecture Notes in Mathematics 3, Springer-Verlag, Berlin, 1964.

Adem, J.

1. *The iteration of Steenrod squares in algebraic topology*, Proc. Natl. Acad. Sci. *38* (1952), 720–726.
2. *Relations on iterated reduced powers*, Proc. Natl. Acad. Sci. *39* (1953), 636–638.

Atiyah, M.F., R. Bott, and A. Shapiro

1. *Clifford modules*, Topology *3* (1964), suppl. 1, 3–38.

Barratt, M. G.

1. *Track groups I, II*, Proc. London Math. Soc. *5* (1955), 71–106, 285–329.
2. *Note on a formula due to Toda*, Jour. London Math. Soc. *36* (1961), 95–96.
3. *Homotopy operations and homotopy groups*, (mimeographed notes), Seattle, 1963.

Barratt, M. G., and P. J. Hilton

1. *On join operations of homotopy groups*, Proc. London Math. Soc. *3* (1953), 430–445.

Blakers, A. L., and W. S. Massey

1. *The homotopy groups of a triad I, II, and III*, Ann. Math. *53* (1951), 161–205; *55* (1952), 192–201; *58* (1953), 409–417.

Bott, R.

1. *The stable homotopy of the classical groups*, Ann. Math. *70* (1959), 313–337.

Bousfield, A., E. B. Curtis, D. M. Kan, D. G. Quillen, D. L. Rector, and J. W. Schlesinger

1. *The mod-p lower central series and the Adams spectral sequence*, Topology *5* (1966), 331–342.

Cartan, H.

1. *Une théorie axiomatique des carrés de Steenrod*, Compt. Rend. Acad. Sci. Paris *230* (1950), 425–427.
2. *Sur les groupes d'Eilenberg-MacLane $H(\pi,n)$*, Proc. Natl. Acad. Sci. *40* (1954), 467–471, 704–707.
3. *Sur l'itération des opérations de Steenrod*, Comm. Math. Helv. *29* (1955), 40–58.
4. Séminaire H. Cartan 1950–1951, Paris.
5. Séminaire H. Cartan 1954–1955, Paris.
6. Séminaire H. Cartan 1958–1959, Paris.

Cartan, H., and S. Eilenberg

1. *Homological algebra*, Princeton Mathematical Series 19, Princeton University Press, Princeton, 1956.

Dold, A.

1. *Über die Steenrodschen Kohomologieoperationen*, Ann. Math. (2) *73* (1961), 258–294.

Eckmann, B.

1. *Gruppentheoretischer Beweis des Satzes von Hurwitz-Radon*, Comm. Math. Helv. *15* (1942), 358–366.

Eilenberg, S.

1. *Cohomology and continuous mappings*, Ann. Math. *41* (1940), 231–251.

Eilenberg, S., and S. MacLane

1. *Relations between homology and homotopy groups of spaces I, II*, Ann. Math. *46* (1945), 480–509; *51* (1950), 514–533.
2. *On the groups $H(\pi,n)$ I, II, and III*, Ann. Math. *58* (1953), 55–106; *60* (1954), 49–139, 513–557.

Fadell, E.

1. *On fibre spaces*, Trans. Am. Math. Soc. *90* (1959), 1–14.

Freudenthal, H.

1. *Über die Klassen der Sphärenabbildungen*, Comp. Math. *5* (1937), 299–314.

Godement, R.

1. *Topologie algébrique et théorie des faisceaux*, Publications de l'Institut de mathématique de l'Université de Strasbourg 13, Hermann, Paris, 1958.

Hilton, P. J.

1. *An introduction to homotopy theory*, Cambridge Tracts in Mathematics and Mathematical Physics 43, Cambridge University Press, Cambridge, 1953.

Hopf, H.

1. *Über die Abbildungen der dreidimensionalen Sphäre auf die Kugelfläche*, Math. Ann. *104* (1931), 637–665.
2. *Über die Abbildungen von Sphären auf Sphären niedrigerer Dimension*, Fundam. Math. *25* (1935), 427–440.

Hu, S.-T.

1. *Homotopy theory*, Pure and Applied Mathematics VIII, Academic Press, New York, 1959.

Husemoller, D.

1. *Fibre bundles*, McGraw-Hill, New York, 1966.

Kahn, D. W.

1. *Induced maps for Postnikov systems*, Trans. Am. Math. Soc. *107* (1963), 432–450.

Kan, D. M.

1. *Semisimplicial spectra*, Illinois J. Math. *7* (1963), 463–478.
2. *On the k-cochains of a spectrum*, Illinois J. Math. *7* (1963), 479–491.

Kan, D. M., and G. W. Whitehead

1. *The reduced join of two spectra*, Topology *3* (1965), suppl. 2, 239–261.

MacLane, S.

1. *Homology*, Springer-Verlag, 1963.

Mahowald, M. E.

1. *On the Adams spectral sequence*, Canad. J. Math. 20 (1968), 252–256.

Mahowald, M. E., and M. C. Tangora

1. *Some differentials in the Adams spectral sequence*, Topology *6* (1967), 349–369.

Massey, W. S.

1. *Exact couples in algebraic topology I, II*, Ann. Math. *56* (1952), 363–396; *57* (1953), 248–286.
2. *Products in exact couples*, Ann. Math. *59* (1954), 558–569.

Maunder, C. R. F.

1. *Cohomology operations of the n-th kind*, Proc. London Math. Soc. *13* (1963), 125–154.
2. *Some differentials in the Adams spectral sequence*, Proc. Cambridge Phil. Soc. *60* (1964), 409–420.

May, J. P.

1. *The cohomology of restricted Lie algebras and of Hopf algebras*, dissertation, Princeton University, 1964.

Meyer, J.-P.

1. *Principal fibrations*, Trans. Am. Math. Soc. *107* (1963), 177–185.

Miller, C. E.

1. *The topology of rotation groups*, Ann. Math. (2) *57* (1953), 90–113.

Milnor, J.

1. *Lectures on characteristic classes* (mimeographed notes), Princeton University, n.d.
2. *The Steenrod algebra and its dual*, Ann. Math. (2) *67* (1958), 150–171.
3. *On spaces having the homotopy type of a CW-complex*, Trans. Am. Math. Soc. *90* (1959), 272–280.

Milnor, J., and J. C. Moore

1. *On the structure of Hopf algebras*, Ann. Math. (2) *81* (1965), 211–264.

Oguchi, K.

1. *A generalization of secondary composition and its application*, J. Fac. Sci., Univ. Tokyo, *10* (1963), 29–79.

Peterson, F. P.

1. *Functional cohomology operations*, Trans. Am. Math. Soc. *86* (1957), 187–197.

Peterson, F. P., and N. Stein

1. *Secondary cohomology operations: two formulas*, Am. J. Math. *81* (1959), 281–305.

Postnikov, M. M.

1. *Determination of the homology groups of a space by means of homotopy invariants* (in Russian), Dokl. Akad. Nauk SSSR, 1951, Tom 76, 3, 359–362.
2. *On the homotopy type of polyhedra* (in Russian), Dokl. Akad. Nauk SSSR, 1951, Tom 76, 6, 789–791.
3. *On the classification of continuous mappings* (in Russian), Dokl. Akad. Nauk SSSR, 1951, Tom 79, 4, 573–576.
4. *Investigations in homotopy theory of continuous mappings*, Am. Math. Soc. Translations, series 2, 7 (1957), 1–134.

Puppe, D.

1. *Homotopiemenge und ihre induzierten Abbildungen, I*, Math. Z. *69* (1958), 299–344.

Serre, J.-P.

1. *Homologie singulière des espaces fibrés*, Ann. Math. (2) *54* (1951), 425–505.
2. *Cohomologie modulo 2 des complexes d'Eilenberg-MacLane*, Comm. Math. Helv. *27* (1953), 198–231.
3. *Groupes d'homotopie et classes de groupes abéliens*, Ann. Math. (2) *58* (1953), 258–294.

Spanier, E. H

1. *Algebraic topology*, McGraw-Hill, New York, 1966.
2. *Secondary operations on mappings and cohomology*, Ann. Math. (2) *75* (1962), 260–282.
3. *Higher order operations*, Trans. Am. Math. Soc. *109* (1963), 509–539.

Steenrod, N. E.

1. *The topology of fibre bundles*, Princeton Mathematical Series 14, Princeton University Press, Princeton, 1951.

2. *Cohomology operations and obstructions to extending continuous functions*, Colloquium lectures, Princeton University, 1957.
3. *Products of cocycles and extensions of mappings*, Ann. Math. (2) *48* (1947), 290–320.
4. *Cohomology invariants of mappings*, Ann. Math. (2) *50* (1949), 954–988.
5. *Homology groups of symmetric groups and reduced power operations*, Proc. Natl. Acad. Sci. *39* (1953), 213–223.
6. *Cohomology operations derived from the symmetric group*, Comm. Math. Helv. *31* (1957), 195–218.

Steenrod, N. E., and D. B. A. Epstein

1. *Cohomology operations*, Annals of Mathematical Studies 50, Princeton University Press, Princeton, 1962.

Steenrod, N. E., and J. H. C. Whitehead

1. *Vector fields on the n-sphere*, Proc. Natl. Acad. Sci. *37* (1951), 58–63.

Tangora, M. C.

1. *The cohomology of the Steenrod algebra*, dissertation, Northwestern University, 1966.

Thom, R.

1. *Espaces fibrés en sphères et carrés de Steenrod*, Ann. Sci. École Norm. Supér. *69* (1952), 109–182.

Toda, H.

1. *Composition methods in homotopy groups of spheres*, Annals of Mathematical Studies 49, Princeton University Press, Princeton, 1962.
2. *p-Primary components of homotopy groups IV, Compositions and toric constructions*, Memoirs University of Kyoto *32* (1959), 297–332.

Wall, C. T. C.

1. *Generators and relations for the Steenrod algebra*, Ann. Math. (2) *72* (1960), 429–444.

Whitehead, G. W.

1. *Homotopy theory*, M.I.T. Press, Cambridge, Mass., 1966.
2. *A generalization of the Hopf invariant*, Ann. Math. *51* (1950), 192–237.
3. *The* $(n + 2)$*nd homotopy group of the n-sphere*, Ann. Math. *52* (1950), 245–247.
4. *On the Freudenthal theorems*, Ann. Math. *57* (1953), 209–228.
5. *Generalized homology theories*, Trans. Am. Math. Soc. *102* (1962), 227–283.

Whitehead, J. H. C.

1. *Combinatorial homotopy I, II*, Bull. Am. Math. Soc. *55* (1949), 213–245, 453–496.
2. *A certain exact sequence*, Ann. Math. *52* (1950), 51–110.

Zeeman, E. C.

1. *A proof of the comparison theorem for spectral sequences*, Proc. Cambridge Phil. Soc. *53* (1957), 57–62.

INDEX